Mark W. McClure · Roland N. Horne

Discrete Fracture Network Modeling of Hydraulic Stimulation

Coupling Flow and Geomechanics

Mark W. McClure
Petroleum and Geosystems Engineering
University of Texas at Austin
Austin, TX
USA

Roland N. Horne
Energy Resources Engineering
Stanford University
Stanford, CA
USA

ISSN 2191-5369   ISSN 2191-5377   (electronic)
ISBN 978-3-319-00382-5   ISBN 978-3-319-00383-2   (eBook)
DOI 10.1007/978-3-319-00383-2
Springer Cham Heidelberg New York Dordrecht London

Library of Congress Control Number: 2013938759

Printed on acid-free paper

Springer is part of Springer Science+Business Media (www.springer.com)

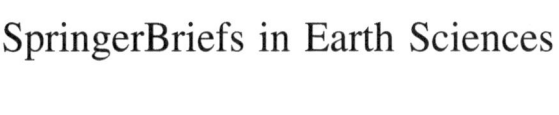
SpringerBriefs in Earth Sciences

For further volumes:
http://www.springer.com/series/8897

Preface

In the past decade, development of unconventional oil and gas resources has been perhaps the biggest new trend in the energy sector. It has become possible to produce huge new resources that were previously considered uneconomic. Key enabling technologies have been long horizontal wells and multiple fracturing stages (King 2010).

Hydraulic fracturing in geothermal energy is typically referred to as Enhanced Geothermal Systems, or EGS (Tester 2007). EGS researchers hope that improvements in stimulation design will enable geothermal to one day experience the same growth in production that is now being seen in unconventional oil and gas.

Despite years of research and field experience, fundamental questions remain unanswered. What is the relative importance of new and preexisting fractures in stimulation? Why do the fracture networks generated in unconventional oil and gas frequently appear to be so complex? What does a stimulated, unconventional reservoir "look like?"

Computational modeling has the potential to be very useful for stimulation design, helping engineers make routine design decisions, test novel ideas, perform formation evaluation, and make decisions about data collection. A remarkable diversity of models and modeling approaches exists. Unfortunately, progress is hampered by the nonuniqueness caused by limited and incomplete data, resulting in difficulty in confirming model assumptions. Therefore, while stimulation modeling holds tremendous promise, it remains a work in progress. A tremendous amount of work is being done in this area, and progress is expected in the years ahead.

This book summarizes a hydraulic stimulation model developed especially for unconventional applications in oil, natural gas, and geothermal. The model described in this book remains incomplete. Several important physical processes remain neglected or incompletely described. However, the strength of this model is that it fully, efficiently, couples fluid flow and the stresses caused by deformation in large, complex discrete fracture networks. We believe that realistically capturing induced stresses will lead to better insight into stimulation processes. Furthermore, we believe that the complexities of the reservoir—spatially variable fracture density, orientation, stresses, and so on—have profound effects on the stimulation process that cannot easily be simplified out of a numerical model without obscuring important details.

Several novel techniques were developed for this work, and existing methods were combined in novel ways. We hope that this work provides a useful contribution to the field of stimulation modeling and lays the groundwork for future research.

References

King, G.: Thirty years of gas shale fracturing: what have we learned? SPE 133456, paper presented at the SPE annual technical conference and exhibition. Florence (2010). doi: 10.2118/133456-MS

Tester, J. (ed.): The Future of Geothermal Energy: Impact of Enhanced Geothermal Systems (EGS) on the United States in the 21st Century. Massachusetts Institute of Technology (2007). http://geothermal.inel.gov/publications/future_of_geothermal_energy.pdf

Acknowledgments

This work was supported financially by the Precourt Institute for Energy at Stanford University. There were many people at Stanford and elsewhere who were very kind to share their expertise as this work was performed. We want to especially thank Drs. David Pollard and Andrew Bradley. Dr. Pollard provided the code COMP2DD that was used to verify the accuracy of the model. Dr. Bradley shared his code Hmmvp that is used by CFRAC for efficient matrix multiplication. Thank you to Jakob Alejandre, a Stanford undergraduate who helped with code development as part of a summer research project.

Contents

Chapter 1
Introduction

Computational modeling of hydraulic fracturing can be used for stimulation optimization, long term production forecasting, basic research into fundamental mechanisms, and investigation of novel techniques. However, modelers face daunting challenges: a variety of complex physical processes, limited and imperfect information, and a heterogeneous and discontinuous spatial domain. Because of these difficulties, it is necessary to make trade-offs between efficiency, spatial resolution, and inclusion of physical processes.

The classical conceptual model of hydraulic fracturing is that a single opening-mode fracture propagates away from the wellbore (Khristianovich and Zheltov 1959; Perkins and Kern 1961; Geertsma and de Klerk 1969; Nordgren 1972). This conceptual model is still used in hydraulic fracturing design and modeling in conventional settings (Economides and Nolte 2000; Meyer and Associates 2011; Adachi et al. 2007). However, in settings with complex fracture networks, classical hydraulic fracturing models may be overly simplified.

This book describes the development and testing of a two-dimensional hydraulic stimulation model that has been developed specifically for simulating hydraulic fracturing in settings with very low matrix permeability and where preexisting fractures play an important role, such as Enhanced Geothermal Systems (EGS) or gas shale. The model simulates fluid flow in a discrete fracture network (DFN), directly discretizing individual fractures (e.g., Karimi-Fard et al. 2004; Golder Associates 2009). The DFN approach is distinct from effective continuum modeling, which averages fracture properties into effective properties over volumetric grid blocks (Warren and Root 1963; Kazemi et al. 1976; Lemonnier and Bourbiaux 2010). The model calculates the stresses induced by opening and sliding along individual fractures and couples opening and slip to fracture transmissivity. The model couples deformation with flow using iterative coupling (Kim et al. 2011) and is efficient enough to simulate problems involving thousands of individual fractures in a few hours or days on a single processor. The model assumes flow is single-phase and isothermal, matrix permeability is negligible, and deformation is small strain in an infinite, linearly elastic medium. The model was written using C++.

M. W. McClure and R. N. Horne, *Discrete Fracture Network Modeling of Hydraulic Stimulation*, SpringerBriefs in Earth Sciences, DOI: 10.1007/978-3-319-00383-2_1, © The Author(s) 2013

The model has several other specific limitations. The model can simulate propagation of new tensile fractures, but the locations of the potentially forming new fractures must be specified in advance (Sects. 2.3.8 and 2.4.1). In practice, a large number of potentially forming fractures can be specified throughout the model, which allows the simulator significant freedom in determining the overall propagation of stimulation (Sects. 2.4.1, 3.3.2, and 4.3). The model sometimes encounters problems at low-angle fracture intersections ($<20°$) as a consequence of its use of the boundary element method to calculate stresses induced by deformation. In practice, this limitation can be handled either by not permitting models to contain low-angle intersections (Sect. 2.4.1) or by implementing a special penalty method (Sects. 2.5.6, 2.3.4, and 2.4.3). These solutions are imperfect, and research is ongoing to identify other solutions.

An earlier version of the model (McClure and Horne 2011) was used to study induced seismicity by performing rate and state friction earthquake simulation. That ability is retained in the current model, but rate and state friction simulations are very intensive computationally and are not discussed in this book. Most simulations described in this book were performed with a constant coefficient of friction. As a compromise between rate and state friction and constant coefficient of friction, static/dynamic friction was implemented and tested in the work described in this book (Sects. 2.5.3, 3.3.1, 4.2.5). Static/dynamic friction is a more efficient, but less rigorous, way of modeling seismicity than rate and state friction.

Earlier versions of the model used in this book have been used to investigate shear stimulation and induced seismicity in fracture networks (McClure and Horne 2010a, b). The model described in this book has significant improvements on the earlier versions. The most important differences are that the model in this book is more efficient, handles fracture opening much more realistically, and is fully convergent to discretization refinement in time.

The philosophy of development has been to construct a model that is realistic and efficient enough to simulate the most important physical processes that take place during hydraulic stimulation in large, complex fracture networks in very low permeability rock. In some cases, special treatments are used in the model to enforce realistic behavior in the fracture network. Our guiding principle is that a model should include as much realism as needed to reasonably capture the physical processes that significantly affect the results of the model for a given purpose.

Because the boundary element method is used, temperature is constant, and matrix permeability is zero, it is only necessary to discretize the fractures, not the volume around the fractures. As a result, the problem size is reduced dramatically. With iterative and approximate methods, it is possible to solve the large, dense systems of equations associated with the use of the boundary element method efficiently in a coupled, implicit fluid flow and geomechanical simulator.

The book is divided into several parts. In Sects. 1.1 and 1.2, the literature is reviewed to identify other numerical models that have been used for complex hydraulic fracturing. In the Methodology (Chap. 2), the model is described in detail. Topics include governing and constitutive equations, numerical methods for mechanical and fluid flow calculations, the coupling scheme between the mechanical and flow equations, handling of mechanical inequality constraints and changing stress boundary conditions, stochastic generation of discrete fracture networks, spatial discretization of fracture networks, and several specialized topics particular to the model, including efficient matrix multiplication, adaptive domain adjustment, handling of regions near opening crack tips, and a technique for handling difficulties associated with low-angle fracture intersections.

In Chaps. 3 and 4, the results of simulations on four different fracture networks are presented and discussed. These simulations (1) verify the accuracy of the mechanical calculations, (2) verify convergence from discretization refinement in both time and space, (3) test sensitivity to error tolerances, and (4) test the effect of various special simulation options, including the use of dynamic/static friction for seismicity modeling, ignoring or including stresses caused by normal displacement of "closed" fractures (in this book, the word "open" is used to refer to a fracture whose walls have mechanically separated because fluid pressure exceeds normal stress, and the word "closed" is used to refer to a fracture with walls that are in contact), and an adjustment to handle low-angle fracture intersections.

The primary purpose of simulations described in this book was to test the accuracy, efficiency, and versatility of the model and to identify optimal simulation parameters. In addition, there are some interesting physical insights that can be drawn from the results. The results demonstrate that stresses induced by deformation affect the propagation of stimulation profoundly, justifying the considerable effort required to integrate them fully into a discrete fracture network simulator.

In McClure (2012), there is discussion of the variety of conceptual models that have been used in the literature to describe the complex fracturing mechanisms taking place in EGS and gas shale. These models involve either opening and propagation of new fractures, sliding of preexisting fractures, or combinations of both. At a given site, establishing the correct conceptual model should be considered a fundamental requirement for successful application of modeling and reservoir engineering.

1.1 Discrete Fracture Network Modeling

In discrete fracture network (DFN) modeling, the equations of fluid flow are solved on each individual fracture. Figure 1.1 shows an example of a discrete fracture network (the network is Model B, discussed in Sects. 3.3.1 and 4.2). The network is two-dimensional, and the fractures should be considered vertical, strike-slip fractures viewed in plan view (equivalently, they could be considered normal fractures viewed in cross-section).

There are several reasons why DFN models are particularly useful for describing complex hydraulic fracturing in low permeability medium. In low permeability rock, individual fractures that are in close proximity but not touching are not well connected hydraulically. As a result, fluid flow between locations is dependent on idiosyncrasies of the fracture network geometry. Connectivity can be

Fig. 1.1 Example of a discrete fracture network. The *blue lines* are preexisting fractures and the thick *black line* represents the wellbore

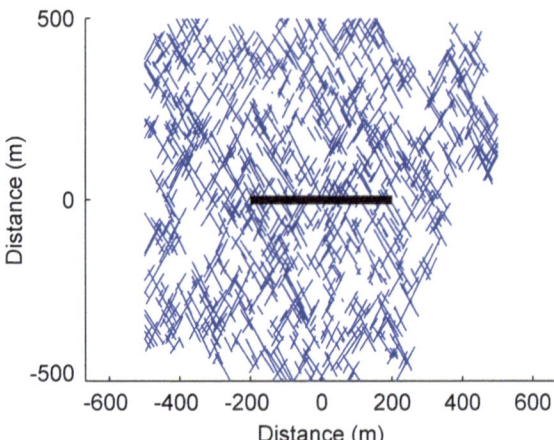

highly unpredictable, with weak connections between two nearby locations, but strong connection between more distant locations (McCabe et al. 1983; Cacas et al. 1990; Abelin et al. 1991). In hydraulically stimulated, fractured, low matrix permeability reservoirs, production logs typically show that productivity can be highly variable along individual wells within the same rock unit (Baria et al. 2004; Dezayes et al. 2010; Miller et al. 2011). Discrete fracture models are better at describing this sort of highly channelized, unpredictable behavior.

Stress perturbation can be handled much more accurately with a DFN. The stresses caused by fracture opening or sliding are very heterogeneous spatially, and effects on neighboring fractures are dependent on their relative orientations and locations. Stresses caused by fracture deformation have major effects on the process of stimulation, as described in this book (see Sect. 4.2.6 for an example of what happens when induced stresses are neglected) and also in McClure and Horne (2010a).

1.2 Review of Stimulation Models

There is a remarkable diversity of hydraulic stimulation modeling described in the literature. The diversity reflects uncertainty about mechanisms of stimulation in the subsurface, the difficulty of validating the underlying assumptions of a given model, the great number of numerical methods available, and different choices about balancing realism and efficiency. This section reviews models of hydraulic fracturing found in the literature. Models that treat hydraulic stimulation as a single planar fracture propagating away from a wellbore are used widely in the industry and in academia (Adachi et al. 2007; Meyer and Associates 2011). However, this section focuses on reviewing models intended for more complex settings.

In some modeling of unconventional fracturing, effective continuum modeling has been used (Yamamoto 1997; Taron and Elsworth 2009; Vassilellis et al. 2011; Lee and Ghassemi 2011; Kelkar et al. 2012). These models average flow from multiple fractures into effective continuum properties. Other models use hybrid approaches that combine aspects of continuum models with aspects of discrete fracture models (Xu et al. 2010; Meyer and Bazan 2011; Palmer et al. 2007).

Some discrete fracture models neglect the stresses caused by deformation of fractures. Neglecting stress interaction reduces the complexity of the implementation significantly and increases efficiency. With this approach, it is possible to perform simulations on large, reservoir-scale discrete fracture networks. In some of these models, fluid flow is calculated by upscaling the DFN to an effective continuum model (Lanyon et al. 1993; Fakcharoenphol et al. 2012; Tezuka et al. 2005; Jing et al. 2000; Willis-Richards et al. 1996; Kohl and Mégel 2007; Rahman et al. 2002; Cladouhos et al. 2011; Wang and Ghassemi 2012; Du et al. 2011). In other models, discrete fracture network are used directly without upscaling to effective continuum models (Bruel 1995, 2007; Sausse et al. 2008; Dershowitz et al. 2010).

A variety of numerical methods have been used to calculate stresses induced by deformation in discrete fracture networks. These methods include finite element (Heuze et al. 1990; Swenson and Hardeman 1997; Rahman and Rahman 2009; Lee and Ghassemi 2011; Fu et al. 2012; Kelkar et al. 2012), finite difference (Hicks et al. 1996; Taron and Elsworth 2009; Roussel and Sharma 2011), boundary element (Asgian 1989; Zhang and Jeffrey 2006; Cheng 2009; Olson and Dahi-Taleghani 2009; Zhou and Ghassemi 2011; Weng et al. 2011; Meyer and Bazan 2011; Safari and Ghassemi 2011; Jeffrey et al. 2012; Tarasovs and Ghassemi 2012; Sesetty and Ghassemi 2012), block-spring model (Baisch et al. 2010), extended finite element (Dahi-Taleghani and Olson 2011; Keshavarzi and Mohammadi 2012), distinct element method (Pine and Cundall 1985; Last and Harper 1990; Rachez and Gentier 2010; Nagel et al. 2011), hybrid finite element/discrete element (Rogers et al. 2010), and the particle-based distinct element method (Suk Min and Ghassemi 2010; Deng et al. 2011; Zhao and Young 2011; Damjanac et al. 2010). An overview of the numerical methods used in rock mechanics (including all of these methods) can be found in Jing et al. (2003).

The majority of published results using coupled geomechanical-fluid flow models in discrete fracture networks have involved a limited number of fractures. Some studies have investigated induced slip on one or a few preexisting fractures (Baisch et al. 2010; Zhou and Ghassemi 2011; Safari and Ghassemi 2011). Some have investigated one or a few propagating and preexisting fractures (Heuze et al. 1990; Zhang and Jeffrey 2006; Rahman and Rahman 2009; Suk Min and Ghassemi 2010; Deng et al. 2011; Zhao and Young 2011; Dahi-Taleghani and Olson 2011; Jeffrey et al. 2012; Sesetty and Ghassemi 2012; Cheng 2009; Keshavarzi and Mohammadi 2012). Some have investigated induced slip on a network of preexisting fractures (no propagation of new fractures) of up to a few dozen fractures (Pine and Cundall 1985; Asgian 1989; Last and Harper 1990; Hicks et al. 1996; Swenson and Hardeman 1997; Rachez and Gentier 2010). Some have investigated propagation of multiple fractures with no preexisting fractures (Tarasovs and Ghassemi 2012; Roussel and Sharma 2011). Some have combined propagation of new fractures and preexisting fractures on networks up to dozens of fractures (Damjanac et al. 2010; Nagel et al. 2011, 2012; Fu et al. 2012, Weng et al. 2011; Rogers et al. 2010).

As the size and complexity of the fracture network increases, the challenge of geomechanical discrete fracture modeling grows considerably. Finite element and finite difference methods require discretization of the area (in two dimensions) or the volume (in three dimensions) around the fractures, and this can lead to a very large number of elements for complex networks and arbitrary fracture geometries. Boundary element methods (BEM) avoid the need to discretize around fractures, but if they are applied directly, they are inefficient for very large problems because they require solution of dense matrices. That disadvantage is mitigated in the model described in this book because iterative and approximate methods are used to improve efficiency of the BEM solutions. Other disadvantages are that the BEM typically cannot handle arbitrary rock properties and heterogeneity and struggles to handle low angle fracture intersections accurately (as discussed in Sect. 3.4 and 4.4). Extended finite element methods may eventually become a powerful technique for hydraulic fracturing modeling but are relatively new and have not yet been demonstrated on complex networks.

Particle-based distinct element models have several advantages. They are versatile because they have the ability to handle arbitrary rock properties and they can avoid discretization difficulties by employing a loose discretization of the volume around fractures. However, in these models, macroscopic rock properties such as Young's modulus are emergent properties of the model, and model inputs (including the coarseness of the discretization itself) must be tuned with trial and error to find settings that match desired behavior (Potyondy and Cundall 2004).

References

Abelin, H., Birgersson, L., Moreno, L., Widén, H., Ågren, T., Neretnieks, I.: A large-scale flow and tracer experiment in granite: 2 results and interpretation. Water Resour. Res. **27**(12), 3119–3135 (1991). doi:10.1029/91WR01404

Adachi, J., Siebrits, E., Peirce, A., Desroches, J.: Computer simulation of hydraulic fractures. Int. J. Rock Mech. Min. Sci. **44**(5), 739–757 (2007). doi:10.1016/j.ijrmms.2006.11.006

Asgian, M.: A numerical model of fluid-flow in deformable naturally fractured rock masses. Int. J. Rock Mech. Min. Sci. Geomech. Abstr. **26**(3–4), 317–328 (1989). doi:10.1016/0148-9062(89)91980-3

Baisch, S., Vörös, R., Rothert, E., Stang, H., Jung, R., Schellschmidt, R.: A numerical model for fluid injection induced seismicity at Soultz-sous-Forêts. Int. J. Rock Mech. Min. Sci. **47**(3), 405–413 (2010). doi:10.1016/j.ijrmms.2009.10.001

Baria, R., Michelet, S., Baumgärtner, J., Dyer, B., Gerard, A., Nicholls, J., Hettkamp, T., Teza, D., Soma, N., Asanuma, H., Garnish, J., and Megel, T.: Microseismic monitoring of the world's largest potential HDR reservoir. Paper presented at the twenty-ninth workshop on geothermal reservoir engineering, Stanford University, https://pangea.stanford.edu/ERE/db/IGAstandard/record_detail.php?id=1702 (2004)

Bruel, D.: Heat extraction modelling from forced fluid flow through stimulated fractured rock masses: application to the Rosemanowes Hot Dry Rock reservoir. Geothermics **24**(3), 361–374 (1995). doi:10.1016/0375-6505(95)00014-H

Bruel, D.: Using the migration of the induced seismicity as a constraint for fractured Hot Dry Rock reservoir modelling. Int. J. Rock Mech. Min. Sci. **44**(8), 1106–1117 (2007). doi:10.1016/j.ijrmms.2007.07.001

Cacas, M.C., Ledoux, E., Marsily, G.D., Barbreau, A., Calmels, P., Gaillard, B., Margritta, R.: Modeling fracture flow with a stochastic discrete fracture network: calibration and validation: 2 the transport model. Water Resour. Res. **26**(3), 491–500 (1990). doi:10.1029/WR026i003p00491

Cheng, Y.: Boundary element analysis of the stress distribution around multiple fractures: implications for the spacing of perforation clusters of hydraulically fractured horizontal wells, SPE 125769. Paper presented at the SPE eastern regional meeting, Charleston (2009). doi:10.2118/125769-MS

Cladouhos, T.T., Clyne, M., Nichols, M., Petty, S., Osborn, W.L., Nofziger, L.: Newberry Volcano EGS demonstration stimulation modeling. Geoth. Resour Counc. Trans. **35**, 317–322 (2011)

Dahi-Taleghani, A., Olson, J.: Numerical modeling of multistranded-hydraulic-fracture propagation: accounting for the interaction between induced and natural fractures. SPE J. **16**(3), 575–581 (2011). doi:10.2118/124884-PA

Damjanac, B., Gil, I., Pierce, M., and Sanchez, M.: A new approach to hydraulic fracturing modeling in naturally fractured reservoirs, ARMA 10-400. Paper presented at the 44th U.S. Rock mechanics symposium and 5th U.S.-Canada Rock mechanics symposium, Salt Lake City, Utah (2010)

Deng, S., Podgorney, R., and Huang, H.: Discrete element modeling of rock deformation, fracture network development and permeability evolution under hydraulic stimulation. Paper presented at the thirty-sixth workshop on geothermal reservoir engineering, Stanford University, http://www.geothermal-energy.org/pdf/IGAstandard/SGW/2011/deng.pdf? (2011)

Dershowitz, W.S., Cottrell, M.G., Lim, D.H., and Doe, T.W.: A discrete fracture network approach for evaluation of hydraulic fracture stimulation of naturally fractured reservoirs, ARMA 10-475. Paper presented at the 44th U.S. Rock mechanics symposium and 5th U.S.-Canada Rock mechanics symposium, Salt Lake City (2010)

Dezayes, C., Genter, A., Valley, B.: Structure of the low permeable naturally fractured geothermal reservoir at Soultz. C. R. Geosci. **342**(7–8), 517–530 (2010). doi:10.1016/j.crte.2009.10.002

Du, C., Zhan, L., Li, J., Zhang, X., Church, S., Tushingham, K., and Hay, B.: Generalization of dual-porosity-system representation and reservoir simulation of hydraulic fracturing-stimulated shale gas reservoirs, SPE 146534. Paper presented at the SPE annual technical conference and exhibition, Denver (2011). doi:10.2118/146534-MS

Economides, M.J., Nolte, K.G.: Reservoir Stimulation, 3rd edn. John Wiley, New York (2000)

Fakcharoenphol, P., Hu, L., and Wu, Y.: Fully-implicit flow and geomechanics model: application for enhanced geothermal reservoir simulations. Paper presented at the thirty-seventh workshop on geothermal reservoir engineering, Stanford University, https://pangea.stanford.edu/ERE/db/IGAstandard/record_detail.php?id=8254 (2012)

Fu, P., Johnson, S.M., Carrigan, C.R.: An explicitly coupled hydro-geomechanical model for simulating hydraulic fracturing in arbitrary discrete fracture networks. Int. J. Numer. Anal. Meth. Geomech. (2012). doi:10.1002/nag.2135

Geertsma, J., de Klerk, F.: A rapid method of predicting width and extent of hydraulically induced fractures. J. Petrol. Technol. **21**(12), 1571–1581 (1969). doi:10.2118/2458-PA

Golder Associates: User documentation: FracMan 7: interactive discrete feature data analysis, geometric modeling, and exploration simulation (2009)

Heuze, F.E., Shaffer, R.J., Ingraffea, A.R., Nilson, R.H.: Propagation of fluid-driven fractures in jointed rock. Part 1—development and validation of methods of analysis. Int. J. Rock Mech. Min. Sci. Geomech. Abstr. 27(4), 243–254 (1990). doi:10.1016/0148-9062(90)90527-9

Hicks, T.W., Pine, R.J., Willis-Richards, J., Xu, S., Jupe, A.J., Rodrigues, N.E.V.: A hydro-thermo-mechanical numerical model for HDR geothermal reservoir evaluation. Int. J. Rock Mech. Min. Sci. Geomech. Abstr. 33(5), 499–511 (1996). doi:10.1016/0148-9062(96)00002-2

Jeffrey, R., Wu, B., Zhang, X.: The effect of thermoelastic stress change in the near wellbore region on hydraulic fracture growth. Paper presented at the thirty-seventh workshop on geothermal reservoir engineering, Stanford University, https://pangea.stanford.edu/ERE/db/IGAstandard/record_detail.php?id=8287 (2012)

Jing, Z., Willis-Richards, J., Watanabe, K., and Hashida, T.: A three-dimensional stochastic rock mechanics model of engineered geothermal systems in fractured crystalline rock. J. Geophys. Res. 105(B10), 23663–23679 (2000). doi:10.1029/2000JB900202

Jing, L.: A review of techniques, advances and outstanding issues in numerical modelling for rock mechanics and rock engineering, Int. J. Rock Mechanics and Mining Sci. 40(3), 283–353 (2003). doi:10.1016/S1365-1609(03)00013-3

Karimi-Fard, M., Durlofsky, L.J., Aziz, K.: An efficient discrete-fracture model applicable for general-purpose reservoir simulators. SPE J. 9(2), 227–236 (2004). doi:10.2118/88812-PA

Kazemi, H., L.S, M., Porterfield, K.L., Zeman, P.R.: Numerical simulation of water-oil flow in naturally fractured reservoirs. Soc. Petrol. Eng. J. 16(6), 317–326 (1976). doi:10.2118/5719-PA

Kelkar, S., Lewis, K., Hickman, S., Davatzes, N.C., Moos, D., and Zyvoloski, G.: Modeling coupled thermal-hydrological-mechanical processes during shear stimulation of an EGS well. Paper presented at the thirty-seventh workshop on geothermal reservoir engineering, Stanford University, https://pangea.stanford.edu/ERE/db/IGAstandard/record_detail.php?id=8295 (2012)

Keshavarzi, R., and Mohammadi, S.: A new approach for numerical modeling of hydraulic fracture propagation in naturally fractured reservoirs, SPE 152509. Paper presented at the SPE/EAGE European unconventional resources conference and exhibition, Vienna, Austria (2012). doi:10.2118/152509-MS

Khristianovich, S.A., and Zheltov, Y.P.: Theoretical principles of hydraulic fracturing of oil strata. Paper presented at the fifth world petroleum congress, New York (1959)

Kim, J., Tchelepi, H., Juanes, R.: Stability, accuracy, and efficiency of sequential methods for coupled flow and geomechanics. SPE J. 16(2), 249–262 (2011). doi:10.2118/119084-PA

Kohl, T., Mégel, T.: Predictive modeling of reservoir response to hydraulic stimulations at the European EGS site Soultz-sous-Forêts. Int. J. Rock Mech. Min. Sci. 44(8), 1118–1131 (2007). doi:10.1016/j.ijrmms.2007.07.022

Lanyon, G.W., Batchelor, A.S., and Ledingham, P.: Results from a discrete fracture network model of a Hot Dry Rock system. Paper presented at the eighteenth workshop on geothermal reservoir engineering, Stanford University, https://pangea.stanford.edu/ERE/db/IGAstandard/record_detail.php?id=2245 (1993)

Last, N.C., and Harper, T.R.: Response of fractured rock subject to fluid injection. Part I. Development of a numerical model. Tectonophysics 172(1–2), 1–31 (1990). doi:10.1016/0040-1951(90)90056-E

Lee, S.H., and Ghassemi, A.: Three-dimensional thermo-poro-mechanical modeling of reservoir stimulation and induced seismicity in geothermal reservoir. Paper presented at the thirty-sixth workshop on geothermal reservoir engineering, Stanford University, https://pangea.stanford.edu/ERE/db/IGAstandard/record_detail.php?id=7261 (2011)

Lemonnier, P., and Bourbiaux, B.: Simulation of naturally fractured reservoirs. State of the art. Oil Gas Sci. Technol.—*Revue de l'Institut Français du Pétrole* 65(2), 239–262 (2010). doi:10.2516/ogst/2009066

McCabe, W.J., Barry, B.J., Manning, M.R.: Radioactive tracers in geothermal underground water flow studies. Geothermics 12(2–3), 83–110 (1983). doi:10.1016/0375-6505(83)90020-2

McClure, M.W.: Modeling and characterization of hydraulic stimulation and induced seismicity in geothermal and shale gas reservoirs. Stanford University, Stanford (2012)

McClure, M.W., and Horne, R.N.: Discrete fracture modeling of hydraulic stimulation in enhanced geothermal systems. Paper presented at the thirty-fifth workshop on geothermal reservoir engineering, Stanford University, https://pangea.stanford.edu/ERE/db/IGAstandard/record_detail.php?id=5675 (2010a)

McClure, M.W., Horne, R.N.: Numerical and analytical modeling of the mechanisms of induced seismicity during fluid injection. Geoth. Resour. Counc. Trans. **34**, 381–396 (2010b)

McClure, M.W., and Horne, R.N.: Investigation of injection-induced seismicity using a coupled fluid flow and rate/state friction model. Geophysics **76**(6), WC181–WC198 (2011) doi:10.1190/geo2011-0064.1

Meyer and Associates, I.: Meyer fracturing simulators, 9th ed. http://www.mfrac.com/documentation.html (2011)

Meyer, B., and Bazan, L.: A discrete fracture network model for hydraulically induced fractures: theory, parametric and case studies, SPE 140514. Paper presented at the SPE hydraulic fracturing technology conference, The Woodlands, Texas, USA (2011). doi:10.2118/140514-MS

Miller, C., Waters, G., and Rylander, E.: Evaluation of production log data from horizontal wells drilled in organic shales, SPE 144326. Paper presented at the North American Unconventional Gas Conference and Exhibition, The Woodlands, Texas, USA (2011). doi:10.2118/144326-MS

Nagel, N., Gil, I., Sanchez-Nagel, M., and Damjanac, B.: Simulating hydraulic fracturing in real fractured rocks - overcoming the limits of pseudo3d models, SPE 140480. Paper presented at the SPE hydraulic fracturing technology conference, The Woodlands, Texas, USA (2011). doi:10.2118/140480-MS

Nordgren, R.P.: Propagation of a vertical hydraulic fracture. Soc. Petrol. Eng. J. **12**(4), 306–314 (1972). doi:10.2118/3009-PA

Olson, J., and Dahi-Taleghani, A.: Modeling simultaneous growth of multiple hydraulic fractures and their interaction with natural fractures, SPE 119739. Paper presented at the SPE hydraulic fracturing technology conference, The Woodlands, Texas (2009). doi:10.2118/119739-MS

Palmer, I., Moschovidis, Z., and Cameron, J.: Modeling shear failure and stimulation of the Barnett Shale after hydraulic fracturing, SPE 106113. Paper presented at the SPE hydraulic fracturing technology conference, College Station, Texas (2007). doi:10.2118/106113-MS

Perkins, T.K., Kern, L.R.: Widths of hydraulic fractures. J. Petrol. Technol. **13**(9), 937–949 (1961). doi:10.2118/89-PA

Pine, R.J., and Cundall, P.A.: Applications of the fluid-rock interaction program (FRIP) to the modelling of hot dry rock geothermal energy systems. Paper presented at the international symposium on fundamentals of rock joints, Bjorkliden (1985)

Potyondy, D.O., Cundall, P.A.: A bonded-particle model for rock. Int. J. Rock Mech. Min. Sci. **41**(8), 1329–1364 (2004). doi:10.1016/j.ijrmms.2004.09.011

Rachez, X., and Gentier, S.: 3D-hydromechanical behavior of a stimulated fractured rock mass. Paper presented at the world geothermal congress, Bali. http://www.geothermal-energy.org/pdf/IGAstandard/WGC/2010/3152.pdf (2010)

Rahman, M., Rahman, M.: A fully coupled numerical poroelastic model to investigate interaction between induced hydraulic fracture and pre existing natural fracture in a naturally fractured reservoir: potential application in tight gas and geothermal reservoirs, SPE 124269. Paper presented at the SPE annual technical conference and exhibition, New Orleans, Louisiana (2009). doi:10.2118/124269-MS

Rahman, M.K., Hossain, M.M., Rahman, S.S.: A shear-dilation-based model for evaluation of hydraulically stimulated naturally fractured reservoirs. Int. J. Numer. Anal. Meth. Geomech. **26**(5), 469–497 (2002). doi:10.1002/nag.208

Rogers, S., Elmo, D., Dunphy, R., Bearinger, D.: Understanding hydraulic fracture geometry and interactions in the Horn River Basin through DFN and numerical modeling, SPE 137488.

Paper presented at the Canadian unconventional resources and international petroleum conference, Calgary, Alberta, Canada (2010). doi:10.2118/137488-MS

Roussel, N., Sharma, M.: Strategies to minimize frac spacing and stimulate natural fractures in horizontal completions, SPE 146104. Paper presented at the SPE annual technical conference and exhibition, Denver (2011). doi:10.2118/146104-MS

Safari, M.R., Ghassemi, A.: 3D analysis of huff and puff and injection tests in geothermal reservoirs, paper presented at the thirty-sixth workshop on geothermal reservoir engineering Stanford University, https://pangea.stanford.edu/ERE/db/IGAstandard/record_detail.php?id=7217 (2011)

Sausse, J., Dezayes, C., Genter, A., Bisset, A.: Characterization of fracture connectivity and fluid flow pathways derived from geological interpretation and 3D modelling of the deep seated EGS reservoir of Soultz (France). Paper presented at the thirty-third workshop on geothermal reservoir engineering, Stanford University, https://pangea.stanford.edu/ERE/db/IGAstandard/record_detail.php?id=5270 (2008)

Sesetty, V., Ghassemi, A.: Modeling and analysis of stimulation for fracture network generation. Paper presented at the thirty-seventh workshop on geothermal reservoir engineering, Stanford University, https://pangea.stanford.edu/ERE/db/IGAstandard/record_detail.php?id=8356 (2012)

Suk Min, K., Zhang, Z., Ghassemi, A.: Hydraulic fracturing propagation in heterogeneous rock using the VMIB method. Paper presented at the thirty-fifth workshop on geothermal reservoir engineering Stanford University, https://pangea.stanford.edu/ERE/db/IGAstandard/record_detail.php?id=7229 (2010)

Swenson, D., Hardeman, B.: The effects of thermal deformation on flow in a jointed geothermal reservoir. Int. J. Rock Mech. Min. Sci., **34**(3–4), 308.e301–308.e320 (1997). doi:10.1016/S1365-1609(97)00285-2

Tarasovs, S., Ghassemi, A.: On the role of thermal stress in reservoir stimulation. Paper presented at the thirty-seventh workshop on geothermal reservoir engineering Stanford University, https://pangea.stanford.edu/ERE/db/IGAstandard/record_detail.php?id=8372 (2012)

Taron, J., Elsworth, D.: Thermal–hydrologic–mechanical–chemical processes in the evolution of engineered geothermal reservoirs. Int. J. Rock Mech. Min. Sci. **46**(5), 855–864 (2009). doi:10.1016/j.ijrmms.2009.01.007

Tezuka, K., Tamagawa, T., Watanabe, K.: Numerical simulation of hydraulic shearing in fractured reservoir. Paper presented at the world geothermal congress, Antalya, Turkey, https://pangea.stanford.edu/ERE/db/IGAstandard/record_detail.php?id=1000 (2005)

Vassilellis, G., Bust, V., Li, C., Cade, R., Moos, D.: Shale engineering application: the MAL-145 project in West Virginia, SPE 146912. Paper presented at the Canadian unconventional resources conference, Alberta, Canada (2011). doi:10.2118/146912-MS

Wang, X., Ghassemi, A.: A 3D thermal-poroelastic model for geothermal reservoir stimulation. Paper presented at the thirty-seventh workshop on geothermal reservoir engineering, Stanford University, https://pangea.stanford.edu/ERE/db/IGAstandard/record_detail.php?id=8382 (2012)

Warren, J.E., Root, P.J.: The behavior of naturally fractured reservoirs. Soc. Petrol. Eng. J. **3**(3), 245–255 (1963). doi:10.2118/426-PA

Weng, X., Kresse, O., Cohen, C.-E., Wu, R., Gu, H.: Modeling of hydraulic-fracture-network propagation in a naturally fractured formation. SPE Prod. Oper. **26**(4), 368–380 (2011). doi:10.2118/140253-PA

Willis-Richards, J., Watanabe, K., Takahashi, H.: Progress toward a stochastic rock mechanics model of engineered geothermal systems. J. Geophys Res. **101**(B8), 17481–17496 (1996). doi:10.1029/96JB00882

Xu, W., Thiercelin, M., Ganguly, U., Weng, X., Gu, H., Onda, H., Sun, J., Le Calvez: Wiremesh: a novel shale fracturing simulator, SPE 132218. Paper presented at the international oil and gas conference and exhibition, Beijing (2010). doi:10.2118/132218-MS

Yamamoto, T., Kitano K., Fujimitsu Y., Ohnishi H.: Application of simulation code, GEOTH3D, on the Ogachi HDR site, paper presented at the 22nd annual workshop on geothermal reservoir engineering, Stanford University, https://pangea.stanford.edu/ERE/db/IGAstandard/record_detail.php?id=463 (1997)

Zhang, X., Jeffrey, R.G.: The role of friction and secondary flaws on deflection and re-initiation of hydraulic fractures at orthogonal pre-existing fractures. Geophys. J. Int. **166**(3), 1454–1465 (2006). doi:10.1111/j.1365-246X.2006.03062.x

Zhao, X., Young R.P.: Numerical modeling of seismicity induced by fluid injection in naturally fractured reservoirs. Geophysics **76**(6), WC167–WC180 (2011). doi:10.1190/geo2011-0025.1

Zhou, X., Ghassemi, A.: Three-dimensional poroelastic analysis of a pressurized natural fracture. Int. J. Rock Mech. Min. Sci. **48**(4), 527–534 (2011). doi:10.1016/j.ijrmms.2011.02.002

Chapter 2
Methodology

The model described in this book computes fluid flow and deformation in discrete fracture networks. As an input, the model requires a realization of the preexisting fracture network. The model has the ability to describe propagation of new fractures, but the potential locations of new fractures must be specified in advance.

Because the boundary element method is used and fluid flow in the matrix surrounding the fractures is assumed negligible, it is not necessary to discretize the volume around the fractures, significantly reducing the number of elements in the discretization.

In the following sections, the modeling methodology is given in detail, including the governing and constitutive equations, numerical methods of solution, methods of discretization, methods of generating realizations of the preexisting (and potentially forming) discrete fracture network, and several special simulation techniques used for realism and efficiency.

2.1 Governing and Constitutive Equations

In the model, fluid flow is single-phase and isothermal. The unsteady-state fluid mass balance equation in a fracture is (adapted from Aziz and Settari 1979):

$$\frac{\partial(\rho E)}{\partial t} = \nabla \cdot (q_{flux} e) + s_a, \tag{2.1}$$

where s_a is a source term (mass per time for a unit of area of the fracture), t is time, E is void aperture (the pore volume per unit area of the fracture), ρ is fluid density, q_{flux} is mass flux (mass flow rate per cross-sectional area of flow), and e is the hydraulic aperture (the effective aperture for flow in the fracture).

Darcy flow is assumed, in which mass flux in a direction x_i is (Aziz and Settari 1979):

M. W. McClure and R. N. Horne, *Discrete Fracture Network Modeling*
of Hydraulic Stimulation, SpringerBriefs in Earth Sciences,
DOI: 10.1007/978-3-319-00383-2_2, © The Author(s) 2013

$$q_{flux,i} = \frac{k\rho}{\mu_l}\frac{\partial P}{\partial x_i},\tag{2.2}$$

where P is fluid pressure, μ_l is fluid viscosity, and k is permeability.

Fluid density and viscosity are functions of pressure (because the simulations were isothermal) and are interpolated from a polynomial curve fit based on values generated using the Matlab script XSteam 2.6 by Holmgren (2007) assuming a constant temperature of 200 °C. A high temperature is used because of the application to geothermal energy. The choice of higher temperature causes fluid density to be modestly lower and viscosity to be nearly a factor of ten lower, relative to room temperature.

The cubic law for fracture transmissivity (the product of permeability and hydraulic aperture) is (Jaeger et al. 2007):

$$T = ke = \frac{e^3}{12}.\tag{2.3}$$

Hydraulic aperture is equal to void aperture between two smooth plates but is lower than void aperture for rough surfaces such as rock fractures (Liu 2005). A "fracture" in a DFN model may represent a crack, but it may also represent a more complex feature such as a fault zone. In the latter case, the void aperture may be much larger than the hydraulic aperture. The model allows e and E to be different.

For single-phase flow in a one-dimensional fracture, the mass flow rate q is:

$$q = \frac{Th\rho}{\mu_l}\frac{\partial P}{\partial x},\tag{2.4}$$

where h is the "out-of-plane" dimension of the flowing fracture (for example, the height of a vertical fracture that is viewed as a one-dimensional fracture in plan view).

Fluid flow boundary conditions (representing the wellbore) are either constant rate or constant pressure wellbore boundary conditions (Sect. 2.3.10). The boundaries of the spatial domain are impermeable to flow.

Stresses induced by deformation are calculated according to the equations of quasistatic equilibrium in a continuum assuming that body forces are equal to zero. These stresses are given by the vector equation (Jaeger et al. 2007):

$$\nabla^T \mathbf{T}_s = 0,\tag{2.5}$$

where $\mathbf{T}_s$ is the stress tensor.

Linear elasticity in an isotropic, homogeneous body is assumed, giving the following relationship between stress and strain according to Hooke's law (Jaeger et al. 2007):

$$\mathbf{T}_s = \frac{2Gv_p}{1-2v_p}trace(\varepsilon)I + 2G\varepsilon,\tag{2.6}$$

where I is the unit matrix, ε is the strain tensor, v_p is Poisson's ratio, and G is the shear modulus.

The shear cumulative displacement discontinuity at any point, D, is equal to the time integral of sliding velocity, v:

$$D = \int v dt. \qquad (2.7)$$

A distinction is made between mechanically open and closed fractures. An open fracture is in tension such that the walls are physically separated and out of contact. A closed fracture bears compressive stress, and its walls are in contact.

For a closed fracture, Coulomb's law requires that shear stress be less than or equal to the frictional resistance to slip. We include an additional term, $v\eta$ (the radiation damping term), to approximate the effect of inertia during sliding at high velocities (Rice 1993; Segall 2010). The radiation damping coefficient, η, is defined to be equal to $G/(2v_s)$, where v_s is the shear wave velocity (Rice 1993; Segall 2010). The radiation damping term is on the order of several MPa, which means that the radiation damping term is small unless sliding velocity is at least centimeters per second. The Coulomb failure criterion with a radiation damping term is (Jaeger et al. 2007; Segall 2010):

$$|\tau - \eta v| = \mu_f \sigma'_n + S_0, \qquad (2.8)$$

where μ_f is the coefficient of friction, S_0 is fracture cohesion, and σ'_n is the effective normal stress, defined as (Jaeger et al. 2007):

$$\sigma'_n = \sigma_n - P, \qquad (2.9)$$

where compressive stresses are taken to be positive. For fractures with shear stress less than the frictional resistance to slip, shear deformation is assumed to be negligible.

Force balance requires that the effective normal stress of open fractures is zero. Because the fluid inside open fractures cannot support shear stress, the walls are stress free (Crouch and Starfield 1983). These stress conditions can be stated:

$$\sigma'_n = 0, \qquad (2.10)$$

$$\tau - \eta v = 0. \qquad (2.11)$$

Relationships are used to relate effective normal stress and cumulative shear displacement to void and hydraulic aperture. These relationships are chosen to be consistent with laboratory derived relations and so that there is not any discontinuity as elements transition between being open and closed. The aperture of a closed fracture is defined as (Willis-Richards et al. 1996; Rahman et al. 2002; Kohl and Mégel 2007):

$$E = \frac{E_0}{1 + 9\sigma'_n/\sigma_{n,Eref}} + D_{E,eff} \tan \frac{\varphi_{Edil}}{1 + 9\sigma'_n/\sigma_{n,Eref}}, \tag{2.12}$$

where E_0, $\sigma_{n,Eref}$, and φ_{Edil} are specified constants. $D_{E,eff}$ is defined as equal to D if $D < D_{E,eff,max}$, and equal to $D_{E,eff,max}$ otherwise. The constants are allowed to be different for hydraulic aperture, e, and void aperture E. Non-zero φ_{Edil} corresponds to pore volume dilation with slip, and non-zero φ_{edil} corresponds to transmissivity enhancement with slip.

The void and hydraulic aperture of an open preexisting fracture is defined as:

$$E = E_0 + D_{E,eff} \tan \varphi_{Edil} + E_{open}, \tag{2.13}$$

$$e = e_0 + D_{e,eff} \tan \varphi_{edil} + E_{open}, \tag{2.14}$$

where E_{open} is the physical separation between the fracture walls.

The hydraulic and void apertures of newly formed fractures are treated differently than preexisting fractures. A value E_{hfres} is defined as the residual aperture of a newly formed fracture. Hydraulic aperture, e, is set equal to void aperture, E. The aperture of an open, newly formed fracture is:

$$E = E_{hfres} + E_{open}, \tag{2.15}$$

and the aperture of a closed, newly formed fracture is:

$$E = E_{hfres} \exp(-\sigma_n K_{hf}), \tag{2.16}$$

where K_{hf} is a specified stiffness for closed hydraulic fracture elements.

If closed, the transmissivity of a newly formed fracture is defined as:

$$T = T_{hf.fac} E_{hfres}, \tag{2.17}$$

and if open, transmissivity is defined as:

$$T = T_{hf.fac} E_{hfres} + (E_{open})^3 / 12, \tag{2.18}$$

where $T_{hf.fac}$ is a specified constant. This treatment of transmissivity for newly forming fractures is used so that, if desired, they can be assigned a relatively high residual transmissivity. This might be desirable as a very simple way of approximating the effect of proppant in newly formed fractures, which would tend to cause higher residual transmissivity after closure (Fredd et al. 2001).

2.2 Initial Conditions

In the results shown here, the fluid pressure and stress state (defined by P, σ_{xx}, σ_{yy}, and σ_{xy}) were assumed to be constant everywhere in the model at the beginning of a simulation. In natural systems, the stress state or pore pressure may be spatially heterogeneous. For example, varying mechanical properties between layers can lead to significant stress heterogeneity and is believed to be responsible for vertical confinement of fractures (Warpinski et al. 1982; Teufel and Clark 1984). If desired, spatially variable stress state could be incorporated into the model simply by changing the stress state of individual elements prior to initiating the simulation.

2.3 Methods of Solution

In this section, the numerical methods of solution for the governing equations are described. The simulator solves the unsteady-state mass balance equation, Equation 2.1, while simultaneously satisfying constitutive relations, updating stresses due to deformation, and satisfying the mechanical equations of force equilibrium. The fluid flow equations are solved using the finite volume method and the mechanical deformation problem is solved using the boundary element method of Shou and Crouch (1995). Time stepping is performed with the implicit Euler method, in which all equations and all unknowns are solved simultaneously in a coupled system of equations during every time step (Aziz and Settari 1979).

2.3.1 Iterative Coupling

During each time step, the simulator solves for the primary variables: pressure, P, open aperture, E, and shear displacement, D, based on equations governing conservation of mass and momentum (details in Sects. 2.3.4, 2.3.5, 2.3.6, and 2.3.7). Iterative coupling is used (Kim et al. 2011) to solve the system of equations. Iterative coupling is a strategy for solving a system of equations in which the system is split into parts and then solved sequentially until convergence. Figure 2.1 is a diagram of the iterative coupling approach.

At the beginning of each time step, the shear stresses equations (Eqs. 2.8 and 2.11; or equivalently, 2.31 and 2.32) are solved as a system of equations with sliding displacements as unknowns while pressure and normal displacements are held constant (Sect. 2.3.5). Next the mass balance and normal stress equations (Eqs. 2.1, 2.10, and 2.12; or equivalently, 2.24, 2.26, 2.27) are solved with pressure and normal displacements as unknowns holding shear displacements fixed (Sect. 2.3.4). The shear stress residuals change during the solution to the flow and normal stress equations, and so after solving the flow and normal stress equations, shear stress residuals are rechecked. Convergence occurs if the shear stress residuals are below a certain threshold, *itertol*. The simulator iterates until convergence. In the simulations in this book, convergence typically occurred in less than ten iterations (most commonly around three), with more iterations required for longer time steps.

Fig. 2.1 Summary of the
iterative coupling approach
for a single time step

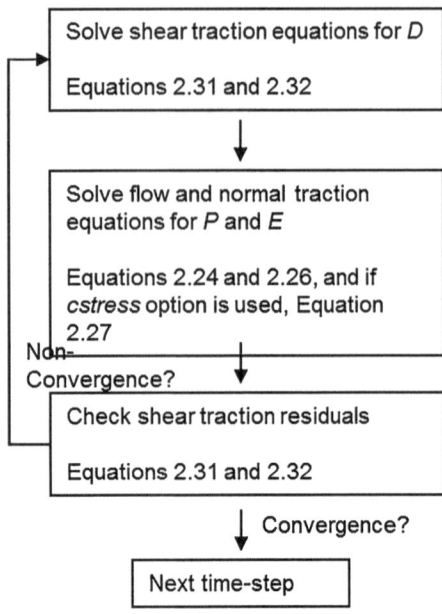

2.3.2 Fracture Deformation: Displacement Discontinuity Method

Stresses induced by deformation are solved with the boundary element method for fracture deformation, the Displacement Discontinuity (DD) Method. Quadratic basis functions are used according to the method of Shou and Crouch (1995). The method of Shou and Crouch (1995) calculates the stresses induced by shear and normal displacement discontinuities by satisfying the equations of quasistatic equilibrium and compatibility for small strain deformation in an infinite, two-dimensional, homogeneous, isotropic, linearly elastic medium in plane strain. The problem reduces to finding the induced stresses $\Delta\tau$ and $\Delta\sigma$ at each element i caused by the shear displacements ΔD and opening displacements ΔE of each element j. Stresses and displacements are linearly related so that:

$$\Delta\sigma_{n,i} = \sum_{j=1}^{n} (B_{E,\sigma})_{ij}\Delta E_j, \tag{2.19}$$

$$\Delta\sigma_{n,i} = \sum_{j=1}^{n} (B_{D,\sigma})_{ij}\Delta D_j, \tag{2.20}$$

$$\Delta\tau_i = \sum_{j=1}^{n} (B_{E,\tau})_{ij}\Delta E_j, \tag{2.21}$$

$$\Delta\tau_i = \sum_{j=1}^{n} (B_{D,\tau})_{ij}\Delta D_j, \tag{2.22}$$

where $B_{E,\sigma}$, $B_{D,\sigma}$, $B_{E,\tau}$, and $B_{D,\tau}$ are matrices of interaction coefficients calculated according to Shou and Crouch (1995). Interaction coefficients do not change during a simulation, and so they can be calculated once and then stored in memory.

The method of Shou and Crouch (1995) assumes strain plane deformation, which implies that the thickness of the deforming medium is infinite in the out-of-plane dimension. In three dimensions, the spatial extent of stress perturbation caused by a deforming fracture scales with the smallest dimension of the fracture (either width or length). Therefore, an implication of plane strain is that the spatial extent of stresses induced by fracture deformation scales linearly with the length of the fracture. In layered formations (more common in gas shale than in EGS), hydraulic fractures are typically confined to mechanical layers and can have length significantly greater than height. In this case, plane strain is not a good assumption. The Olson (2004) adjustment factor can be used to account for the effect of fracture height. Every interaction coefficient is multiplied by a factor:

$$G_{adj,ij} = 1 - \frac{d_{ij}^{\beta}}{(d_{ij}^2 + (h/\alpha)^2)^{\beta/2}}, \tag{2.23}$$

where h is the specified formation height, d is the distance between the centers of the two elements i and j, α is an empirical value equal to one, and β is an empirical value equal to 2.3 (Olson 2004). If h is very large, then the adjustment factor is nearly one, and the method reduces to plane strain.

2.3.3 Stresses Induced by Normal Displacements of Closed Fractures

The simulator has the option to include (the *cstress* option) or not include (the *nocstress* option) the stresses induced by normal displacements of closed fractures.

The *nocstress* method can be justified because closed cracks experience very small opening displacements in response to significant changes in normal stress. From Barton et al. (1985), closed joints are most compliant at normal stresses below 10 MPa. Reducing normal stress from 10 to 0 MPa leads to a normal displacement of 0.1–0.25 mm, depending on the stiffness of the joint. If a closed, 10 m long fracture experiences 0.1 mm of opening displacement, the approximate stress change (assuming plane strain) due to that displacement is less than 0.1 MPa (from Eq. 2.46). Induced stresses would be even less for larger fractures because fracture stiffness is inversely proportional to size (Eq. 2.46). As a result, the stresses induced by opening displacement of closed elements can be neglected with minor effect on the results unless the joints are very small or unusually compliant.

On the other hand, the *cstress* method may be useful because not all closed fractures are actually thin joints. For example, fault zones may be treated as fractures in a DFN, but in reality they can be centimeters to meters thick and be composed of a complex zone of porous material and fracturing (Wibberley et al. 2008; Faulkner et al. 2010). In this case, the fractures represented in a DFN are not literally crack-like features. The void and hydraulic apertures are effective values that include the total effect of all sources of fluid storage and transmissivity in the fault zone. Because fault zones can have very high effective void aperture (if they are thick, porous zones), they can contain a very significant amount of fluid, and as a result induce significantly greater displacement, strain, and stress than if they were thin cracks.

While the overall approach of the *cstress* option could be useful for modeling thick fault zones, there are some problems with the current implementation. The model calculates stresses induced by fracture normal displacements using the Shou and Crouch (1995) Displacement Discontinuity Method. This method is intended to calculate the stresses induced by opening of a crack—a displacement discontinuity in which a solid is parted and physically separated along a discrete plane. The physical process of fluid filling a porous, finite thickness fault zone is obviously different from the opening of a crack. One difference is that for the same amount of injected fluid, stresses induced by the opening of a crack are greater because the fluid is emplaced into a much thinner zone, resulting in a much greater strain (because strain is the ratio of displacement and size). Therefore it is not correct to model poroelastic swelling of a fault zone as the opening of a crack. This problem could be resolved in future work by replacing the Shou and Crouch (1995) method with a BEM designed for volumetric, poroelastic strain. Changing the BEM would affect the value of the interaction coefficients between elements, but the overall numerical approach of the *cstress* method (Sect. 2.3.3) would be unchanged.

2.3.4 Solution to the Fluid Flow and Normal Stress Equations

The fractures are discretized into discrete elements (Sect. 2.4.2). The unsteady-state mass balance equation, Equation 2.1, is solved implicitly using the finite volume method, which leads to a nonlinear system of equations that must be solved at each time step. Equations representing conditions for normal stress are included in the fluid flow system of equations.

With the *cstress* method, there is a $(2n) \times (2n)$ system of equations (n is the total number of elements), with pressure and void aperture of each element as unknowns. With the *nocstress* method, there is a $(n + m) \times (n + m)$ system of equations (m is the number of open elements), with pressure of each element and void aperture of each open element as unknowns. In many simulations, m is much smaller than n, and so the *cstress* method can be significantly more computationally intensive.

From the finite volume method, a mass balance residual equation (from Eqs. 2.1 and 2.4) is written for each individual element m:

$$
R_{m,mass} = \sum_{q=1}^{Q} T_{g,qm}^{n+1} \left(\frac{\rho}{\mu_l}\right)_{qm}^{n+1} \left(P_q^{n+1} - P_m^{n+1}\right)
$$
$$
+ s_m^{n+1} - 2a_m h \frac{(E\rho)_m^{n+1} - (E\rho)_m^n}{dt^{n+1}},
$$

(2.24)

where the superscript n denotes the previous time step, and superscript $n+1$ denotes the current time step, Q is the number of elements connected to element m, q is a dummy subscript referring to each of the elements connecting to element m, a is the element half-length, T_g is the geometric part of the transmissibility between two elements (referred to as geometric transmissibility, see Eq. 2.25 for a definition), s_m is a source term such as a well (units of mass per time; positive s is flow into an element), and dt is the duration of the time step. The equations are solved fully implicitly by evaluating all terms at the upcoming time step. An exception is that when using the *cstress* option, the geometric transmissibility term is evaluated explicitly. The term $(\rho/\mu_l)_{qm}$ is evaluated with upstream weighting.

In this dissertation, the term "transmissivity" is used to refer to the ability of fluid to flow through a fracture (in general), and the term "transmissibility" is used to refer to the ability for fluid to flow between two numerical elements. The geometric transmissibilities are calculated using harmonic averaging. For a connection between adjacent elements not at a fracture intersection, geometric transmissibility is calculated as:

$$
T_{g,qm} = \frac{2h\left(e_q^3/a_q\right)\left(e_m^3/a_m\right)}{12\left(e_q^3/a_q\right) + \left(e_m^3/a_m\right)}.
$$

(2.25)

Geometric transmissibilities at fracture intersections are calculated according to the method of Karimi-Fard et al. (2004), which gives a way to calculate geometric transmissibilities between elements at fracture intersections without requiring a zero-dimensional element at the center of the intersection. In the Karimi-Fard et al. (2004) method, geometric transmissibility between two elements at a fracture intersection depends on all elements at the intersection.

For open elements (with either *nocstress* or *cstress*), the additional residual equation is (from Eq. 2.10):

$$
R_{E,o} = \sigma_n',
$$

(2.26)

and the additional unknown is E_{open} (or equivalently, E).

With the *cstress* method, fracture aperture is an additional unknown (and leads to an additional equation) for each closed element. For closed elements with the *cstress* method, the additional residual equation is (from Eq. 2.12):

$$R_{E,c} = E - \frac{E_0}{1 + 9\sigma'_n/\sigma_{n,Eref}} - D_{E,eff}\tan\frac{\varphi_{Edil}}{1 + 9\sigma'_n/\sigma_{n,Eref}}, \qquad (2.27)$$

with E as the additional unknown.

With the *nocstress* method, stresses induced by changes in aperture of closed elements are neglected. It is not necessary to include an additional unknown and equation for closed elements (such as Eq. 2.27) because the empirical relations for fracture void and hydraulic aperture (Eq. 2.12) can be substituted directly into the fluid flow residual equation (Eq. 2.24). The hydraulic aperture is treated as a function of P, D, and σ_n, independent of E, and substituted into the fluid flow residual equation in both the *cstress* and the *nocstress* methods.

The system of equations is solved using an iterative method similar to Newton–Raphson iteration. In Newton–Raphson iteration, an iteration matrix is defined such that:

$$J_{ij} = \frac{\partial R_i}{\partial X_j}, \qquad (2.28)$$

where i reflects the row number, j reflects the column number, J_{ij} is an entry in the iteration matrix, R_i is an entry in the residual vector, and X_j is an entry in the vector of unknowns (Aziz and Settari 1979). In Newton–Raphson iteration, the iteration matrix is called the Jacobian. However, we do not use the term Jacobian because, as discussed in this section, the iteration matrix used is not a full Jacobian. The algorithm makes a series of guesses for X until certain convergence criteria are met. For each iteration, X is updated according to:

$$X_{new} = X_{prev} + dX, \qquad (2.29)$$

where:

$$J\,dX = -R. \qquad (2.30)$$

After each update of X, the residual vector is recalculated. Stresses are induced by changes in fracture aperture (changes in aperture of all elements if *cstress* is used, changes in aperture for open elements if *nocstress* is used), and so as a part of the residual update, the stresses at each element are updated based on the changes in aperture. The stress update requires matrix multiplication according to Eqs. 2.19 and 2.21. The two most computationally intensive steps in solving the system of equations are solving Eq. 2.30 and updating the stresses.

In the DD method, normal displacements of each element affect the stress at every other element. As a result, the columns of the full Jacobian matrix corresponding to the normal displacements are dense. Solving a large, dense system of equations, as required by Eq. 2.30, is extremely intensive computationally, and if attempted directly, would severely limit the practical size of the problem that could be solved. To handle this difficulty, an incomplete Jacobian matrix is used as an iteration matrix instead of a full Jacobian matrix.

The iteration matrix is equal to the full Jacobian matrix with interaction coefficients of absolute value below a certain threshold neglected. This approach can effectively solve the system of equations because in plane strain, the magnitude of interaction coefficients decays with the inverse of the square of distance (using the correction of Olson 2004, coefficients decay even faster). As a result, the interaction coefficients of immediate neighbors are much larger than interaction coefficients of distant elements.

The threshold for inclusion in the iteration matrix is set to at 5 % of the value of the self-interaction coefficient (the effect of an element's opening displacement on its own normal stress) for open elements and 30 % for closed elements (if the *cstress* option is used). Using these thresholds, the iteration matrix typically includes five to ten interaction coefficients for each open element and even fewer for each closed element.

Neglecting values in the iteration matrix does not affect the accuracy of the solution; it only decreases the quality of the guesses based on Eq. 2.28. After each dX update is calculated, the stresses caused by the changes in opening displacement are fully calculated according to Eqs. 2.19 and 2.21 without neglecting interaction coefficients.

Using an incomplete Jacobian increases the number of iterations required to reach convergence (relative to using a full Jacobian), but hugely reduces the computational burden of solving Eq. 2.30. Because element interactions are affected mainly by near-neighbors (and time steps are short enough that variables change slightly between time steps), convergence is still possible in a reasonable number of iterations. Typically, the simulator achieves convergence in three or four iterations.

The iteration matrix is an unsymmetric sparse matrix. The publicly available code UMFPACK is used to solve the iteration matrix (Davis 2004a, b; Davis and Duff 1997, 1999). Within UMFPACK, the publicly available codes AMD (Amestoy et al. 1996, 2004; Davis 2006), BLAS (Lawson et al. 1979; Dongarra et al. 1988a, b, 1990a, b), and LAPACK (Anderson et al. 1999) are used.

Several criteria are used to judge convergence in the flow/opening stress equations. The change in the fluid pressure from iteration to another must be less than 0.001 MPa, and the change in opening displacement must be less than 1 micron. A normalized residual vector is calculated by multiplying each stress residual equation by 0.1, each mass balance residual by $dt/(2a_iE_ih\rho_i)$, and (if flow rate is specified) the wellbore residual equation by $1/s_i$. For convergence, the infinity norm of the normalized residual is required to be less than 10^{-4}. The Euclidean norm (scaled by the number of elements) is required to be less than 10^{-5}. If a constant pressure boundary condition is specified, the total flow rate into/out of the system is calculated, and the change in the calculated flow rate between iterations is required to be less than 10^{-5} kg/s.

If the nonlinear solver does not converge within a specified number of iterations, the time step is discarded and repeated with a smaller dt. In testing, it was found that nonconvergence was uncommon.

2.3.5 Solution to the Shear Stress Equations

The shear stress equations are solved with cumulative displacement, D, as the unknown (or equivalently, v or ΔD) holding fluid pressure and normal displacements constant. The sliding velocity v during a time step is equal to ΔD, the change in sliding displacement during the time step, divided by dt, the duration of the time step.

Kim et al. (2011) analyzed several iterative coupling schemes between flow and deformation for stability and convergence. They found that when solving the deformation part of the problem, it was better to hold mass content of each element constant than to hold pressure constant in each element. In this work, pressure was held constant while solving the shear deformation equations. In future work, we will experiment with holding mass content constant, and this may improve convergence rate.

The residual equations for closed and open elements are (from Eqs. 2.8 and 2.11):

$$R_{D,closed} = |\tau - \eta v| - \mu \sigma_n - S_0, \tag{2.31}$$

$$R_{D,open} = |\tau - \eta v| - S_{0,open}, \tag{2.32}$$

The shear stress residual equations are solved as a system of equations. The simulator identifies elements that have negative residual and a sliding velocity of zero and categorizes them as being "locked." The shear stiffness of locked elements is assumed to be infinite, and so locked elements have zero sliding velocity. Because the shear displacements of locked elements are not changing, they are excluded from the system of equations.

To be strictly correct, the cohesion term, $S_{0,open}$, should not be included for open elements because open fractures are not able to bear shear stress. The term is included because it offers a numerical convenience and has little effect on the results (as long as $S_{0,open}$ is small). Without the $S_{0,open}$ term, cohesion vanishes abruptly when closed elements become open elements. This causes very rapid sliding, which forces the simulator to take a large number of very short time steps. Because this process can happen frequently during a simulation (every time an element opens), it can significantly reduce efficiency. Possibly, abrupt loss of cohesion during fracture opening is a realistic process, and perhaps it is a cause of microseismicity. We are not aware of any discussion of this process in the literature. From the point of view of numerical modeling, the process is an inconvenience that increases simulation run-time drastically. The inclusion of the $S_{0,open}$ term prevents this process from occurring.

In the Shou and Crouch (1995) method, shear deformation of an element affects stress at every other element. The result is that the system of equations formed from Eqs. 2.31 and 2.32 is dense. As with the normal stress equations (Sect. 2.3.4), an iterative method similar to Newton–Raphson iteration is used. An iteration matrix is formed from the full Jacobian matrix, but entries with absolute value below a certain threshold are set to zero. All interaction coefficients less than a certain factor, $J_{mech,thresh}$ (in the simulations in this book, 0.01 was used), of an element's self-interaction coefficient are removed from the iteration matrix, resulting in a sparse system of equations. A series of iterations are performed until the infinity norm of the shear stress residuals is less than a prescribed tolerance, *mechtol*. Special complications are discussed in Sects. 2.3.6 and 2.3.7. In the simulations in this book, convergence typically occurred in less than five or ten iterations.

The iteration matrix is an unsymmetric sparse matrix. The publicly available code UMFPACK is used to solve the matrix (Davis 2004a, b; Davis and Duff 1997, 1999).

Compared to solving the system of equations directly, the iterative method radically improves the scaling of computation time with size. Direct solution of a dense matrix scales like n^3, where n is the number of elements. Assuming that solving the iteration matrix is a negligible cost, the iterative method reduces the problem to several matrix multiplications, which scale like n^2 (with direct multiplication). As discussed in Sects. 2.5.1, 3.5, and 4.5, an efficient matrix multiplication technique is used that further reduces the problem scaling to between n and $n\log(n)$.

Because the iteration matrix is a general sparse matrix, solving the matrix may not necessarily be a trivial computational expense for larger problems or for larger values of $J_{mech,thresh}$. If the cost of solving the iteration matrix system became prohibitive (in testing for this book, it did not), a banded iteration matrix could be used. A banded iteration matrix would require more iterations because it would not be able to include interactions between nearby elements in adjacent fractures (which are particularly important at fracture intersections). However, the solution of banded matrices is very efficient, and so solution time of the iteration matrix could be guaranteed to be small.

2.3.6 Inequality Constraints on Fracture Deformations

To enforce realistic behavior, two inequality constraints are imposed for fracture deformations. The walls of open fractures may not interpenetrate ($E_{open} \geq 0$), and a fracture may not slide backwards against the direction of shear stress ($\tau \Delta D \geq 0$). During every iteration of the flow/normal stress subloop, prior to updating stresses caused by the changes in displacement, the model checks each element to see if the applied displacement will violate an inequality constraint. If an applied displacement will result in the violation of an inequality constraint, the applied displacement is adjusted so that equality will be satisfied (resulting in E_{open} and ΔD being equal to zero).

Adjustments to enforce the constraints are made frequently by the simulator. When the iteration matrix is solved, the algorithm is unaware of the constraints, and so whenever an element transitions from open to closed or from sliding to not sliding, the update overshoots zero (E_{open} or ΔD equal to zero), and an adjustment must be applied. These adjustments tend to be small and typically have a minor effect on convergence.

However, because adjustments are nonsmooth perturbations to the residual equations, they have the potential to cause nonconvergence. Convergence problems did not occur for any of the simulations performed for this book, but with testing, it was found that convergence could sometimes be a problem when solving the shear stress residual equations on very complex, dense, and/or poorly discretized fracture networks. If nonconvergence occurs, the simulator automatically reduces time step duration. With sufficient reduction in time step, displacements can always be made small enough to assure convergence. However, frequent time step reduction due to convergence failure leads to poor efficiency and is not desirable for optimal performance.

The shear stress residual equations can fail to converge if a complex cluster of intersecting fractures is poorly discretized. In this case, a group of elements can interact in such a way that the iteration matrix consistently attempts to make updates that violate the constraint. A cycle results as the updates violate the constraint and then are reset. Unlike the normal stress element residual equation (Eq. 2.26), the shear stress residual equation of closed elements (Eq. 2.31) contains both normal and shear stress, which increases the potential for complex interactions between elements.

The tendency to for nonconvergence is affected by the number of elements included in the iteration matrix. With more elements included in the iteration matrix, nonconvergence is more likely because there is greater potential for complex interactions between neighboring elements. However, the system converges more rapidly if more elements are included. Therefore, a trade-off exists between efficiency (including more elements in the iteration matrix, which reduces the number of iterations required) and robustness (including fewer elements in the iteration matrix to prevent the possibility of reverse sliding). Including too many elements in the iteration matrix can lead to reduced efficiency if the time required to solve the iteration matrix becomes nonnegligible compared to the time required to update stresses. The most robust iteration matrix is probably one that has zeros everywhere except the main diagonal, but this matrix may require hundreds of iterations for convergence. For these reasons, the user specified parameter $J_{mech,thresh}$, which determines the number of elements included in the iteration matrix, has an important impact on efficiency and robustness. For the simulations in this book, $J_{mech,thresh}$ equal to 0.01 was used.

In the simulations in this book, convergence failure of the shear stress residual equations did not ever occur. However, convergence failure is possible in poorly discretized networks. If this is happens, the best solution is to refine the discretization. An alternative approach is to use a larger value of $J_{mech,thresh}$.

2.3.7 *Changing Mechanical Boundary Conditions*

As discussed in Sects. 2.3.4 and 2.3.5, different forms of the stress equilibrium equation are solved depending on whether an element is open, sliding, or locked (Eqs. 2.26, 2.27, 2.31, and 2.32). A major issue for the simulator is that element status may change during a time step, and so it is not known in advance which equations to solve for each element. The issue of changing boundary conditions is handled in two ways: frequent checking of element status between iterations and time stepping.

The use of time stepping is advantageous because deformation is always small for a sufficiently short time step. Time step duration plays a direct role in the residual equations: in the accumulation term of Eq. 2.24 and the radiation damping term of Eqs. 2.31 and 2.32. Reducing time step sufficiently always enables the simulator to converge because the residual equations do not contain discontinuities (although there are discontinuities in the derivative of the residual equations).

Typically, time step reduction to enable convergence is not necessary because the combination of iterative methods and frequent checking of element status allows convergence to be achieved in the vast majority of cases. Iterative methods are used in both the flow/normal stress and the shear stress subloops. The effect of these iterative methods is that fractures are deformed gradually as they converge to the solution. Changing element status can be problematic because it creates discontinuities in the derivative of the residual equations that worsen convergence. However, because the equations are solved with iterative methods that apply a series of small deformations, the effect of the discontinuities is dampened and typically convergence can be achieved. The need for gradual deformation is another reason why including too many elements in the iteration matrix (using a $J_{mech,thresh}$ that is too low) could be problematic for robustness.

As these gradual deformations are applied, element status is checked constantly. Elements are checked for opening and closing after every iteration in the flow/normal stress subloop and checked for opening/sliding/locking after every iteration in the shear stress subloop. An exception is that elements that are locked at the beginning of the shear stress subloop are not checked during every iteration of that subloop—they are assumed to remain locked. If dynamic friction weakening is used (Sect. 2.5.3), the status of all elements is checked after every iteration of the shear stress subloop. After the shear stress subloop has converged (before beginning the flow/normal stress subloop), element status is checked for every element, even the previously locked elements. Finally, after the flow/normal stress subloop, before evaluating the overall coupling error for the iterative coupling, the element status is checked a final time.

Convergence problems could hypothetically occur if elements entered a cycle in which status switched back and forth from one iteration to another. In testing, cycles were rarely observed even for large problems that had many elements that are opening, sliding, or locked.

Because time stepping is used, typically the deformation equations are solved when the solution (the deformation at the end of the time step) is quite close to the initial guess (the deformation at the beginning of the time step). However, in Sect. 3.2.1, it is demonstrated that the method described in this book can be used for solving contact problems where the initial guess is not close to the solution.

2.3.8 Formation of New Tensile Fractures

The model has the ability to model the propagation of new tensile fractures, but it requires that the location of potential new tensile fractures be specified in advance. The potential new fractures are discretized prior to simulation. Before potential fracture elements become "real" elements (become "active"), they are considered "inactive," which means that they have zero transmissivity and cannot slide or open. Inactive elements are not included in any of the systems of equations. However, the stress state at inactive elements is updated constantly throughout the simulation. Figures 2.2, 2.3, and 2.4 show examples of fracture networks containing both natural and potentially forming fractures.

The process by which an element is activated is discussed in Sect. 2.3.8. When an element is activated, it is given an aperture equal to E_{hfres} (see Eqs. 2.15 and 2.16), set at 10 microns. Because the E_{hfres} is increased from zero to 10 microns, activation of an element does not strictly conserve mass. However, because E_{hfres} is small, the error is small relative to the total amount of fluid injected during stimulation. E_{hfres} could be made smaller, but practically, elements with very small apertures can be problematic for the simulator. Once activated, an element is never deactivated.

Fig. 2.2 An example of a fracture network with prespecified deterministic potential hydraulic fractures. The *black line* is the (*horizontal*) wellbore; the *blue lines* are preexisting fractures, and the *red lines* are the potentially forming hydraulic fractures

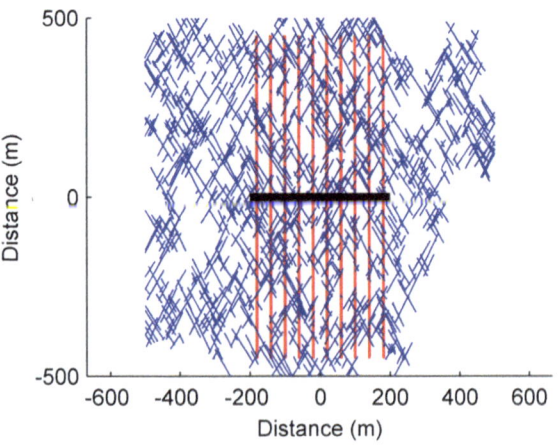

Fig. 2.3 An example of a fracture network with potential hydraulic fractures. The *black line* is the wellbore; the *blue lines* are preexisting fractures, and the *red lines* are the potentially forming hydraulic fractures

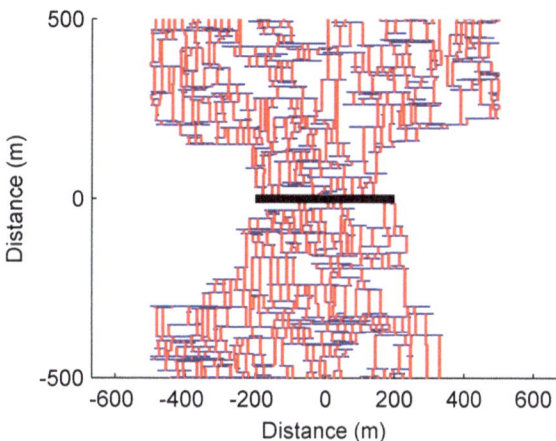

Fig. 2.4 An example of a fracture network with potential hydraulic fractures. The *black line* is the wellbore; the *blue lines* are preexisting fractures, and the *red lines* are the potentially forming hydraulic fractures

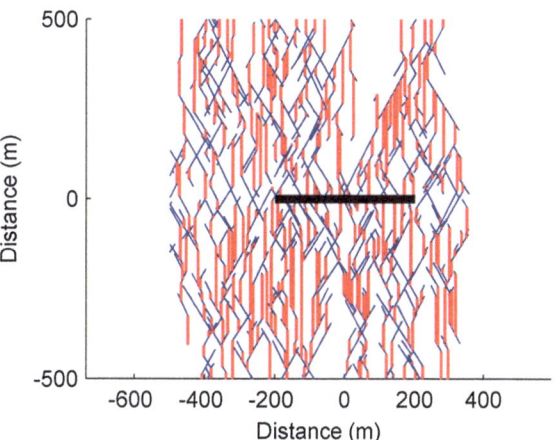

2.3.9 Adaptive Time Stepping

With adaptive time stepping, the duration of each time step is varied in order to optimize efficiency and accuracy. Time steps are selected based on the maximum change in a variable δ_i^n, defined to measure how quickly stress is changing at each element. δ_i^n is defined as the sum of the absolute value of the change in effective normal stress and the absolute value of the change in shear stress at each element i during the time step n. The duration of the next time step is selected using the method suggested by Grabowski et al. (1979):

$$dt^{n+1} = dt^n \min_i \left(\frac{(1+\omega)\eta_{t\,\mathrm{arg}}}{\delta_i^n + \omega\eta_{t\,\mathrm{arg}}} \right), \tag{2.33}$$

where η_{targ} is a user specified target for the maximum change in δ and ω is a factor that can be between zero and one (in this book, ω was set to one). If $\delta_i > 4.0\eta_{targ}$ for any element, the time step is discarded and repeated with a smaller value of dt.

Several secondary conditions are used to select time step duration. If convergence failure occurs in the shear stress residual equations, the flow/opening residual equations, or the overall iterative coupling, the time step duration is cut by 80 % and repeated. The time step duration is sometimes adjusted so that the end of the time step coincides with the prespecified duration of the simulation or a prespecified adjustment in the wellbore boundary conditions (Sect. 2.3.10).

2.3.10 Wellbore Boundary Conditions

Wellbore boundary conditions can be specified by rate or pressure. For injection, the user specifies a maximum injection pressure, P_{injmax}, and a maximum injection rate, q_{injmax}. For production, the user specifies a maximum production rate, $q_{prodmax}$, along with a minimum production pressure, $P_{prodmin}$. A schedule of wellbore control parameters is specified by the user. At each time step, the simulator identifies which constraint should be applied such that the other is not violated and implements the appropriate boundary condition.

For constant rate boundary conditions, the source term s in Eq. 2.24 is nonzero for elements connected to the wellbore. The specified rate is enforced such that it is equal to the sum of the source terms of all elements connected to the wellbore:

$$S = \sum_k s_k, \tag{2.34}$$

where S is the specified injection/production rate (positive for injection) and k cycles through every element connected to the wellbore. Equation 2.34 is added to the system of flow equations, and P_{inj} is the corresponding unknown. The source term of each element connected to the wellbore, s_k is:

$$s_k = \frac{T_{g,wk}(P_{inj} - P_k)\rho_{wk}}{\mu_{l,wk}}. \tag{2.35}$$

The geometric transmissibility between a fracture element and the wellbore, $T_{g,wk}$, is calculated from Eq. 2.25 with the assumption that e_w is infinite (although obviously a wellbore does not literally have a hydraulic aperture.) With a very large e_w, Equation 2.25 reduces to:

$$T_{g,wk} = 2he_k^3/(12a_k). \tag{2.36}$$

If a rate of zero is specified, the wellbore remains connected to the formation and there can be cross-flow between fractures through the wellbore. Pressure drop within the wellbore is assumed negligible. In this case, Equation 2.34 is still included in the flow equations, but S is set to zero.

Constant pressure boundary conditions are implemented by including an element in the flow equations with a very large (effectively infinite) volume and hydraulic aperture. Because the element has a very large volume, its pressure remains constant regardless of fluid flow in or out of the element.

2.4 Spatial Domain

2.4.1 Generation of the Discrete Fracture Network

The fracture network used by the simulator can be generated deterministically or stochastically. If specified deterministically, the user explicitly specifies the locations of the fractures (or the potentially forming fractures).

Simple techniques are used for stochastic fracture network generation. Fractures are generated sequentially with an accept/reject algorithm. The total number of fractures in the network is specified by the user. Before identifying the locations of any of the fractures, a population of lengths and orientations is chosen according to prespecified statistical distributions. The locations of the fractures are then determined in order from largest fracture to smallest. For each fracture, a candidate location is identified, and certain checks are performed before "accepting" the candidate location. If the checks are not satisfied, a new candidate location is selected, and the process is repeated until an acceptable location is found.

Candidate locations are chosen at random within a specified spatial domain. In order to avoid boundary effects, the candidate locations are located in a spatial domain larger than the problem domain. After all the fractures have been generated, the network is cropped to the size of the problem domain.

Candidate fracture locations are accepted or rejected on the basis of several checks designed to avoid numerical difficulties associated with the use of the Displacement Discontinuity Method (Shou and Crouch 1995). As discussed in Sect. 2.5.6, low-angle fracture intersections cause numerical difficulties. To avoid this problem, fractures are not allowed to intersect (or come within one meter of intersecting) at an angle below a specified threshold (in Models B and C in this book, the threshold was 20°). In addition, fracture intersections are not allowed to be closer than one meter.

As discussed in Sect. 2.3.8, new tensile fractures can form, but the potential locations of these new fractures must be specified in advance. There are several simple algorithms that are used to identify the potential locations of these new tensile fractures. In all cases, the potential new hydraulic fractures are assumed to be perpendicular to the remote least principal stress. This is a simplification because, while newly formed fractures should form perpendicular to the least principal stress, deformations during stimulation cause stress perturbation that can rotate the principal stresses locally from its original orientation.

One way to locate the potential new hydraulic fractures is deterministically. To ensure compatibility with the accept/reject requirements for fracture intersections, deterministic potential fractures are located prior to stochastic generation of fractures. An example is shown in Fig. 2.2. The network shown in Fig. 2.2 was used in Model B (Sect. 3.3.1).

Two other algorithms are used for locating potential new tensile fractures. These algorithms are applied after the generation of the stochastic preexisting fracture network (rather than before, as with the deterministic method). These two algorithms can allow propagating fracture to terminate against natural fractures or propagate across natural fractures. In the results shown in this book, the algorithms were used to create fracture networks where propagating fractures terminate against natural fractures.

One algorithm randomly selects a few locations along the wellbore to initiate potential new fractures, then propagates them away from the wellbore until they intersect preexisting fractures. Next, a few random locations along each of the intersected preexisting fractures are chosen, more potential fractures are initiated and propagated, and the process is repeated. A sample discretization using this method is shown in Fig. 2.3. To avoid numerical difficulties, the newly formed fractures are not permitted to be closer than one meter from each other.

Another algorithm locates the potential new fractures off the tips of the preexisting fractures. These fractures are then propagated away from the wellbore until they reach a preexisting fracture, and then the cycle is repeated. An example is shown in Fig. 2.4. The network shown in Fig. 2.4 is the network used for Model C in Sect. 3.3.2.

After all of the fractures have been generated, a breadth-first graph-search algorithm is used to identify which fractures are connected to the wellbore through a continuous pathway through preexisting or potential fractures. Because the matrix permeability is assumed to be zero, fractures that do not have a continuous pathway to the wellbore are hydraulically isolated from the injection. These fractures are omitted from the simulation.

2.4.2 Spatial Discretization

Because the matrix permeability is assumed to be negligible and mechanical deformation is calculated using the boundary element method, it is only necessary to discretize the fractures, not the area around the fractures. An example of a typical discretization is shown in Fig. 2.5. Element size is not required to be constant.

The boundary element method of Shou and Crouch (1995) is inaccurate within a factor (less than one) of an element's half-length. To minimize inaccuracy, the discretization is refined when fractures are close together.

Fig. 2.5 Example of a
fracture network
discretization. Each dot is
located at the center of an
element. Note the refinement
around the fracture
intersections

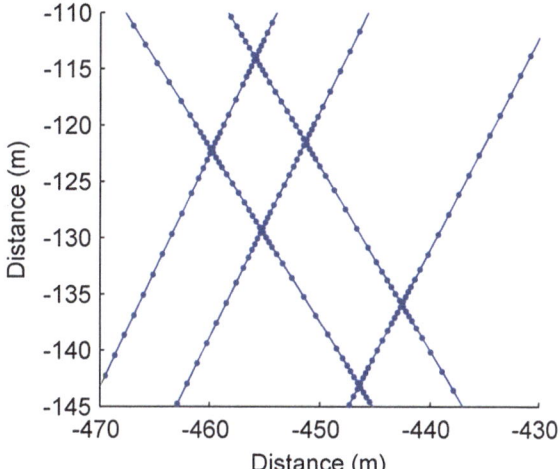

The discretization is performed in two parts. In the initial discretization, the fractures are discretized with a roughly uniform element half-length, a_{const}. In the second phase, the discretization is refined iteratively.

In the initial discretization, the algorithm starts at one end of each fracture and moves across it, generating one element at a time with a length of $2a_{const}$. Fracture intersections cannot occur in the middle of an element, and so an element is terminated if an intersection is reached or the end of the fracture is reached. Termination of elements at intersections or fracture endpoints results in elements that are not $2a_{const}$ long. If a newly created element has a half-length below a certain threshold, $0.5a_{const}$, it is combined with the previous element. An element is not combined with the previous element if the previous element is on the other side of an intersection or if there isn't a previous element because the new element is the first element on the fracture. If the endpoint of a fracture is less than $0.1a_{const}$ from a fracture intersection, the end of the fracture is removed, leaving a "T" shaped intersection.

After generation of the initial discretization, an iterative algorithm is used for further refinement. During each iteration, two conditions are checked for each element, and if an element violates either condition, it is split in half. A minimum element half-length is set, a_{min}, below which an element is not split further. This process is repeated until no further elements are split.

The first condition enforces refinement around fracture intersections:

$$a_i < l_c + d_{ij}l_f, \tag{2.37}$$

where a_i is the half-length of element i, d_{ij} is the distance from the center of element i to intersection j, and l_c and l_f are constants. The second condition prevents elements from being too close together:

$$l_k a_i < D_{ij}, \tag{2.38}$$

where l_k is equal to a specified value, either l_s if elements i and j belong to the same fracture or l_o if i and j belong to different fractures. For elements of the same fracture, D_{ij} is the distance between the center of element j and the edge of element i. For elements of different fractures, D_{ij} is the smallest of three numbers: the distance between the center of element j and either (1) the center of element i or (2) either of the two endpoints of element i.

To avoid evaluating Eq. 2.38 n^2 times every iteration, the spatial domain is divided into a grid, and Eq. 2.38 is only evaluated between elements in neighboring gridblocks or in the same gridblock.

2.5 Special Simulation Topics

2.5.1 Efficient Matrix Multiplication

In the Shou and Crouch (1995) method, deformation at every element affects stress at every other element. As a result, updating stress requires multiplication of a dense matrix of interaction coefficients—a process that scales like n^2. As problem size grows, computation time and RAM requirements quickly become prohibitive.

Fortunately, approximate methods for efficient matrix multiplication are available. Two techniques are the fast multipole method (Morris and Blair 2000) and hierarchical matrix decomposition (Rjasanow and Steinbach 2007). The model in this book uses Hmmvp, a publicly available implementation of hierarchical matrix decomposition and adaptive cross approximation (Bradley 2012). Prior to running simulations, Hmmvp is used to perform matrix decomposition and approximation of the four matrices of interaction coefficients (Eqs. 2.19, 2.20, 2.21, and 2.22) for a given fracture network. The decomposition is stored and loaded into memory at the beginning of a simulation. Within the simulator, Hmmvp is used to perform the stress updating matrix multiplications. In Sects. 3.5 and 4.5, it is demonstrated that Hmmvp drastically reduces the memory and computation requirements for matrix multiplication.

2.5.2 Crack Tip Regions

Special treatment is needed to model the progressive opening of both preexisting and newly forming fractures. As discussed in Sect. 2.3.8, the simulator is capable of modeling the propagation of new tensile fractures, but the locations of these fractures must be specified in advance. The simulator needs to have conditions to handle three situations: (1) the initiation of a new fracture (the activation of the first element on a newly forming fracture), (2) the extension of a newly forming fracture (the activation of subsequent elements on the newly forming fracture), and (3) the progressive opening of a preexisting fracture.

Initiation of new tensile fractures is handled in a very simple way. On each potentially forming hydraulic fracture, the (not yet activated) elements that are touching preexisting fractures or the wellbore are identified and labeled "initiation" elements. Initiation elements are "linked" to their adjoining elements on the neighboring fracture or the wellbore. At the end of each time step, a check is performed on all initiation elements to see if they should be activated. During the check, the fluid pressure of each initiation element is assumed to be equal to the fluid pressure that is highest among the elements that it is linked to. If the effective normal stress of an initiation element is less than zero (tensile), the fracture is initiated. To avoid discretization dependence, all elements on the potentially forming fracture that are within one meter of the intersection (including the initiation element itself) are activated.

Once an element on a potentially forming hydraulic fracture has been activated, a second algorithm is used to extend the fracture. The stress intensity factor at a crack tip is estimated using the Displacement Discontinuity Method according to the equation (Schultz 1988):

$$K_I = \frac{G}{4\pi(1-v)} E_{open} \sqrt{\frac{2\pi}{a}}, \tag{2.39}$$

where E_{open} is the opening of the crack tip element and a is its half-length. If the stress intensity factor reaches a critical threshold, $K_{I,crithf}$, then the fracture is allowed to propagate.

To propagate a fracture, a process zone is defined as the region of the newly forming fracture 2.0 m ahead of the crack tip. If the critical stress intensity factor is reached at the crack tip element, elements within the process zone are activated. For application of Eq. 2.39, a crack tip element is permitted to be any element in the fracture that has been activated. It is not required to be the outermost active element.

A special treatment is also used to handle the propagation of opening along a preexisting fracture. In this case, only part of a preexisting fracture has opened. The location of the transition from open to closed on the fracture is considered an effective crack tip.

Without a special treatment, propagation of opening along a preexisting fracture could be unrealistically slow. In order for opening to propagate along a preexisting fracture, fluid must be able to flow from the open element at the effective crack tip into the adjacent closed element. Because geometric transmissibility between elements is calculated using a harmonic average (Eq. 2.25), the rate of flow between a high transmissivity element and low transmissivity element will be limited by the low transmissivity element. Because of this, without a special adjustment, the propagation of opening down a preexisting fracture will be limited by the transmissivity of the closed elements ahead of the effective crack tip. This is not realistic because the crack tip should be able to propagate at a rate proportional to the (high) transmissivity of the open fracture behind the tip as fluid flows in behind it and progressively opens the crack. This is the same reason why a crack can propagate through very low permeability matrix: propagation depends on fluid flowing in behind the crack tip, not flowing ahead of the crack tip.

Opening induces tensile stresses ahead of the crack tip, but these stresses are unable to open the elements ahead of the effective crack tip because of a poro-elastic response. Because of the low compressibility of water (relative to the fractures) and conservation of mass, effective normal stress of an element must be nearly constant unless there is fluid flow into or out of the element (otherwise the void aperture and resulting mass content of an element would change, from Eqs. 2.12 and 2.24). Because of this effect, the tensile stresses induced ahead of the crack tip produce a reduction in fluid pressure that allows the effective normal stress to remain nearly constant and prevents opening.

To correct for this problem, a special adjustment is used to increase element transmissivity ahead of effective crack tips. The stress intensity factor at an effective crack tip is calculated using Eq. 2.39. If the stress intensity factor reaches a critical value, $K_{I,crit}$ (permitted to be different than $K_{I,crithf}$), then elements within 2.0 m of the effective crack tip (belonging to the same fracture) are placed in a process zone. The hydraulic aperture of the elements in the process zone is assigned a process zone hydraulic aperture, e_{proc} (equal to 106 microns), unless their hydraulic aperture is already higher than e_{proc}. The processes zone hydraulic aperture is high enough that fluid can rather quickly flow into the process zone.

As an aside, it is worth mentioning why a special adjustment is *not* needed for shear stimulation. In shear stimulation, transmissivity enhancement due to induced slip advances along a preexisting fracture. An effective shear crack tip can be defined at the boundary between where slip and transmissivity enhancement have taken place and where they have not. Shear stress concentration occurs ahead of the effective shear crack tip, just as tensile stress concentration occurs ahead of an effective opening crack tip. However, unlike for tensile stresses, there is not a poroelastic response that acts to counteract the effect of the stress concentration. In this case, the induced shear stress can cause shear deformation and transmissivity enhancement ahead of effective crack tip without requiring any fluid to flow ahead of the crack tip, a process we refer to as Crack-like Shear Stimulation (Sects. 3.4.2.2 and 4.4.2 in McClure 2012).

2.5.3 Dynamic Friction Weakening

Seismicity occurs when friction weakens rapidly on a fault, resulting in a rapid release of shear stress through slip. The leading theory to describe friction evolution on a fault is rate and state friction, and it is widely used in earthquake modeling (Dieterich 2007). In rate and state friction, the coefficient of friction changes over time as a function of sliding velocity and the past sliding history of the fracture. According to rate and state friction, the coefficient of friction is (Segall 2010):

$$\mu_f = f_0 + a_{rs} \ln \frac{v}{v_0} + b \ln \frac{\theta v_0}{d_c},$$ (2.40)

where f_0 is a constant of order 0.6–1.0, a_{rs} and b are constants of order 0.01, v_0 is a specified reference velocity, θ is a state variable that changes over time as friction evolves on the fracture, and d_c is a characteristic weakening distance on the order of microns (though in some applications, much larger values are used). In McClure and Horne (2011) of Chap. 1, a numerical approach was demonstrated that couples rate and state friction evolution, deformation, and fluid flow.

Coupling seismicity modeling with fluid flow is useful because in some cases seismicity is triggered by injection or other human activities (McGarr 2002).

Rate and state friction is significantly more intensive computationally than simulations using constant coefficient of friction. One reason rate and state friction models are so computationally intensive is that they are numerically unstable unless a very finely resolved spatial discretization is used (LaPusta 2001). Another reason is that rate and state friction requires a very large number of very short time steps in order to simulate seismic events accurately because frictional weakening is very non-linear.

In McClure and Horne (2011) of Chap. 1, explicit third-order Runge–Kutta time stepping was used. Sliding velocity was calculated by treating velocity as an unknown and calculating the velocity at each element to enforce the frictional equilibrium equation. This approach has significant difficulty when applied in settings where effective normal stress is very low. At low effective normal stress, small perturbations in shear stress can lead to large changes in sliding velocity, making accuracy challenging for explicit time stepping methods unless very small time steps are used. Possibly, implicit time stepping would perform better for settings with low normal stress, but that approach has not been tested.

As an alternative to rate and state friction, static/dynamic friction, was used to compute the results shown in this book. With static/dynamic friction, if the shear stress of an element exceeds the Coulomb failure criterion, Equation 2.8, then the element is considered to be sliding and its coefficient of friction is instantaneously lowered to a new dynamic value, μ_d. The sudden lowering of friction causes rapid sliding and can result in cascading sliding and weakening of friction on neighboring elements. The result is a process that mimics earthquake nucleation and propagation. Once the sliding velocity of a sliding element has gone below a certain threshold, friction is instantaneously restrengthened to its original static value.

When static/dynamic friction is used, checks are performed for frictional weakening and element status for all elements after every iteration in the shear stress subloop (see Sect. 2.3.7). Checks for friction restrengthening are performed at the end of each time step.

Static/dynamic friction has certain drawbacks, but it is a reasonable compromise between realism and efficiency. Static/dynamic friction is tested in Sect. 3.3.1 and discussed in Sect. 4.2.5. In McClure and Horne (2011) of Chap. 1, results from rate and state friction modeling were compared to static/dynamic friction modeling from McClure and Horne (2010b) of Chap. 1, and the results were qualitatively similar.

In McClure and Horne (2010b) of Chap. 1, static/dynamic friction was used, but a radiation damping term was not used, and so all slip during a seismic event was effectively instantaneous and simultaneous. That was a rather problematic approach, and in the present model, the inclusion of a radiation damping term prevents slip from occurring instantaneously (though weakening and restrengthening of friction is instantaneous).

Because static/dynamic friction allows a seismic event to nucleate at a single element (and friction weakens instantaneously), there is an inherent discretization-dependence to the results. Earthquake models with this property have been referred to as "inherently discrete" (Ben-Zion and Rice 1993). In contrast, rate and state simulations that use an adequate spatial discretization require nucleation to occur on a cooperating patch of several elements, and these models are convergent to grid refinement.

A major assumption of the model in this book, quasistatic equilibrium, breaks down at very high slipping velocity, where dynamic stress transfer effects can play an important role. The radiation damping term is used to approximate the effects of dynamic stress transfer, but it is not necessarily accurate for calculations involving more than a single, planar fault (discussed by LaPusta 2001). Full dynamic simulations could be used to solve the problems more accurately, but these are extremely computationally intensive and would not be feasible for large, complex fracture networks.

2.5.4 Alternative Method for Modeling Friction

We have developed an alternative method for earthquake modeling that is intended to replicate results from rate and state friction but with much better efficiency. The method has not been extensively tested, and because it remains a work in progress, results using the method are not given in this book. However, the method is summarized in this section because it is a potential topic for future work.

An ideal method for modeling seismic events should be convergent to grid refinement, give a reasonably accurate answer with a coarse refinement, and give results similar to rate and state friction simulation. There are two particular problems that need to be overcome to make efficient earthquake simulation possible. Problem (1): crack tip discretization error (with a coarse discretization) causes difficulty in modeling rupture propagation. Problem (2): groups of elements must be able to collectively interact to nucleate a rupture.

With crack tip discretization error (Problem 1), the issue is that the stress concentration at the crack tip is underestimated with a coarse discretization. With rate and state friction, friction weakening ahead of a rupture tip occurs because a concentration of stress causes displacement to occur ahead of the tip while the coefficient of friction is still high. If the concentration of stress ahead of the rupture tip is too weak because of inadequate discretization (the initiating rupture patch is discretized by one or a few elements), friction weakening may be artificially limited, and a seismic event may "fizzle out" shortly after nucleation.

The challenge of modeling rupture initiation (Problem 2) is that in order for results to be convergent to grid refinement, multiple elements must be able to cooperate to nucleate a single seismic event. With rate and state friction, there is a characteristic minimum patch size for nucleation of an earthquake (Segall 2010). Therefore, with sufficient discretization refinement, nucleation patches are composed of a large number of elements. Any method where individual elements are able to nucleate ruptures will not be convergent to grid refinement.

Static/dynamic friction does not handle the rupture initiation issue (Problem 2) because a single element is able to nucleate an event, regardless of element size. This happens because in static/dynamic friction the characteristic length scale of friction weakening is zero (an element does not have to displace over any distance in order to experience a drop in friction). As a consequence, the simulation results are discretization dependent.

A strategy called "RSQSim" has been proposed as an efficient method for modeling earthquakes (Dieterich 1995; Dieterich and Richards-Dinger 2010). In this strategy, a seismic event can nucleate from a single element. Semi-analytical treatments of state evolution are used to calculate the timing of earthquake nucleation. RSQSim requires that elements must be larger than the minimum nucleation patch size, and as a result it is not able to model multiple elements collectively nucleating (it does not solve Problem 2), and as with static/dynamic friction, results are discretization dependent. Dieterich (1995) suggested handling the crack tip discretization problem (Problem 1) with a special multiplying factor applied to the elements adjacent to rupture tips. However, it is not clear whether this method is robust because it is effectively a heuristic tuning parameter. Because it is not theoretically derived, it is not clear whether it can be guaranteed to work in all cases. Ideally, it should be possible to derive the crack tip adjustment theoretically without needing to tune it using trial and error. Another issue is that it is unclear how RSQSim could be coupled with fluid flow.

In the following paragraphs, our proposed alternative method for efficient earthquake simulation is described.

The default setting for an element is "stuck." Stuck elements have zero sliding velocity. Stuck elements are converted to "sliding" elements if their shear stress exceeds their frictional resistance to slip according to the Coulomb law (Eq. 2.8) with a constant, static coefficient of friction. Once converted to a sliding element, the element slides according to a velocity strengthening model, equivalent to the rate and state friction expression for friction (Eq. 2.40) with the b coefficient set to zero.

A special condition is applied to sliding elements to determine if they should begin "nucleation." When an element is nucleating, a displacement weakening law is applied such that the coefficient of friction weakens linearly as a function of displacement over a specified displacement distance until it reaches a specified minimum value. Simultaneously, the velocity strengthening term continues to be used. This approach is equivalent to using the rate and state friction law (Eq. 2.40) with a displacement weakening f_0 term and the b coefficient equal to zero.

It can be shown that during nucleation of a rupture using rate and state friction with the aging law for state evolution (Segall 2010), friction evolution is equivalent to linear displacement weakening:

$$\frac{\partial \theta}{\partial t} = 1 - \frac{v\theta}{d_c} \sim -\frac{v\theta}{d_c} \Rightarrow \log(\theta) = \frac{-vt}{d_c} + C = \frac{-D}{d_c} + C \Rightarrow \frac{\partial \mu_f}{\partial D} = -\frac{b}{d_c}, \quad (2.41)$$

where it has been assumed that during rupture, state is decreasing rapidly, and so $|v\theta/d_c| \gg 1.0$.

The minimum value of f_0 (the value at which friction no longer weakens with displacement) can be chosen so that it is equivalent to using a constant f_0 in the rate and state expression with a non-zero b parameter and with the state variable equal to its steady state value for sliding at 1.0 m/s. The steady state value for the state variable can be calculated for a given sliding velocity by setting the time derivative of state equal to zero in the aging law:

$$\frac{\partial \theta}{\partial t} = 1 - \frac{v_{ss}\theta_{ss}}{d_c} = 0 \Rightarrow \theta_{ss} = \frac{d_c}{v_{ss}}. \quad (2.42)$$

It is important to carefully select the condition to determine when an element should begin nucleation (begin to experience displacement weakening). If an element (or a patch of sliding elements) begins to experience displacement weakening, but the patch is smaller than the characteristic minimum patch size, then sliding will relieve shear stress faster than friction weakening induces slip, and a frictional instability (rapid weakening of friction) will not occur. Frictional instability requires that the weakening of friction occurs faster than sliding relieves shear stress. This principle is the basis of our nucleation condition.

The nucleation condition we propose is:

$$\frac{\partial |\tau|}{\partial D} = -\frac{\partial \mu_f}{\partial D} \sigma'_n. \quad (2.43)$$

In Eq. 2.43, the coefficient of friction derivative term on the right-hand side is a constant value because (by definition) weakening is linear with displacement. The left-hand side of Eq. 2.43 is evaluated numerically for each element by determining the actual change in shear stress and displacement experienced by the element during the time step. The shear stress change of an element is affected by its own sliding and also by the sliding of all other elements. If a patch of elements is sliding, the term on the left-hand side of Eq. 2.43 is equivalent to the stiffness of the entire slipping patch. If a single element is sliding, it is equivalent to the stiffness of the single element. If Eq. 2.43 is satisfied, then the slipping patch has become large enough (and its equivalent stiffness low enough) that the initiation of displacement weakening will lead to frictional instability and rapid acceleration of sliding velocity. Testing with well-resolved discretizations found that this method successfully nucleated events when slipping patches reached the same patch size as the full rate and state simulations. The method also worked well with coarse discretizations, even discretizations so coarse that a single element was larger than the minimum nucleation patch size.

A condition is needed for returning sliding elements to the "stuck" state. Recall that if an element is not "stuck," then it is sliding according to a velocity strengthening law (and may experience displacement weakening of friction if it has experience nucleation). If the sliding velocity of an element goes below a certain threshold, it is reset to "locked," and its sliding velocity returns to zero. Different thresholds are used for elements that have nucleated and elements that have not nucleated (the threshold is higher for the nucleated elements). If a nucleated element is returned to the "stuck" state, its coefficient of friction is returned to its initial value (the displacement weakening of friction is reset).

The crack tip discretization error problem is solved by applying a mechanism to enforce "nucleation" at the crack tip. Nucleation is enforced by reducing the f_0 to the minimum value it would reach due to the displacement weakening (which is a specified model parameter derived using Eq. 2.42).

An appropriate condition is needed to determine when to enforce the crack tip adjustment. If nucleation is applied too readily, then ruptures will propagate further than they should. If nucleation is applied too sparingly, ruptures may not propagate far enough. The goal is that simulated ruptures with coarse discretizations will propagate the same distance as if full rate and state simulation were performed with a well resolved discretization. The crack tip adjustment should have no effect if a large number of elements are sliding because the discretization is well resolved and no adjustment is needed. It should also be possible to derive the adjustment in advance, rather than tuning it by trial and error.

It was found that a stress intensity factor approach worked as a method for determining when to apply nucleation at the crack tip. The stress intensity factor (also discussed in Sect. 2.5.2) may be calculated for mode II deformation using the Schultz (1988) method (Eq. 2.39), with sliding displacement D replacing opening displacement E_{open}. With this method, nucleation is enforced at the crack tip if the stress intensity factor exceeds a mode II fracture toughness. It should be possible to derive the fracture toughness from rate and state parameters, but we have not yet done so. We found with trial and error that if an appropriate fracture toughness was used, then the distance of rupture propagation was insensitive to discretization refinement and consistent with a full rate and state simulation.

2.5.5 Adaptive Domain Adjustment

During some simulations, there are large regions of the spatial domain where stresses and fluid pressures change very slowly during all or some of the simulation. Depending on the specifics of the problem, fractures distant from the injector may experience virtually zero fluid flow and be in a stress state such that they are not close to either sliding or opening. Given this situation, there is an opportunity to reduce computational effort by not updating stresses at these elements during every time step and/or by not including them in the flow simulation equations. This strategy is referred to as adaptive domain adjustment. Adaptive domain adjustment is tested in Sects. 3.3.1 and 4.2.4.

Elements that are not deforming and experience virtually zero fluid flow are identified and placed in a *nochecklist*. All other elements are placed in a *checklist*. The stresses on elements in the *nochecklist* are not updated during every time step. Cumulative deformations between *nochecklist* updates are tracked and at intervals the *nochecklist* elements are updated from the cumulative *checklist* deformations. Because linearly elastic deformations are path-independent, updating stresses periodically results in exactly the same final stresses as if they were updated frequently.

As an additional optimization, *nochecklist* elements are removed from the system of fluid flow equations. Removing the *nochecklist* elements from the fluid flow equations causes them to have effectively zero transmissivity.

An algorithm is used to sort elements into the *nochecklist*. To aid in categorization, a separate list called an *activelist* is kept. Once an element is placed in the *activelist*, it remains there for the duration of the simulation. Elements can be added to the *activelist* for the following reasons: connection to the wellbore, sliding residual greater than a specified negative value, *slidetol* (sliding occurs when sliding residual become positive), effective normal stress less than a specified positive value, *opentol* (opening occurs when normal stress becomes negative), or a perturbation from initial fluid pressure greater than 0.1 MPa. In Simulations B5 and B9 (the two simulations in this book that use adaptive domain adjustment), *opentol* was 4.0 and *slidetol* was -2.0. The entire problem domain is gridded into a 30×30 grid, and blocks containing an element in the *activelist* are identified as active blocks. All elements contained in an active block or a block surrounding the active blocks are placed in the *checklist*. All other elements are placed in the *nochecklist*. Overall, the algorithm leads to a region of *checklist* elements around the injector well that spreads out gradually over time as the region of stimulation grows.

2.5.6 Strain Penalty Method

Inaccuracy in the BEM calculations can occur when elements are in close proximity relative to their size (Crouch and Starfield 1983). The best solution to this problem, described in Sect. 2.4.2, is to refine the discretization in locations where fractures are in close proximity. However, this strategy is not practical at low-angle fracture intersections, where neighboring fractures lie very close together over significant distances. Adequately discretizing these geometries requires a large number of extremely small elements, which is numerically undesirable in the fluid flow calculations. To avoid excessive discretization refinement, a minimum element size is specified in the discretization algorithm (Sect. 2.4.2). As discussed in Sect. 2.4.1, one strategy to avoid these difficulties is to avoid creating fracture networks that contain low-angle fracture intersections.

If low-angle intersections are included in a simulation, elements can interact in unstable ways. Displacements and stresses can grow extremely large, and displacement can form in strange, unrealistic patterns (see examples in Sect. 3.4). Not only are these behaviors unrealistic, in severe cases they can cause simulations to be unusable. Some inaccuracy at low-angle fracture intersections is perhaps acceptable, but it is not acceptable for low-angle intersections to cause problems so severe that they prevent a simulation from continuing.

An algorithm, referred to as the strain penalty method, is used to minimize the effects of inaccuracies at low-angle fracture intersections. The algorithm identifies where large strains are beginning to develop and applies penalty stresses to prevent the strains from growing further. This approach could be considered a crude way to mimic rock failure, which is the process in nature that prevents extreme concentrations of stress and strain.

There is no theoretical basis for the strain penalty method. It is intended as a way to prevent catastrophic numerical error. It does not ensure that the solution is completely accurate. With discretization refinement, the region of inaccuracy can be limited to a small region very near the center of the intersection (Sect. 3.4), which minimizes error.

Perfect accuracy at intersections is probably a false goal because in reality, very complex deformations occur at fracture intersections that do not conform to the assumption of small strain, linearly elastic deformation made by the Shou and Crouch (1995) boundary element method. Therefore, numerical accuracy is overwhelmed by error from the model assumptions. This is not an issue unique to our model. Fracture intersections are very challenging for all numerical methods to describe, not just the boundary element method.

In most simulations in this book—Models A, B, and C, the strain penalty method was not used (it was not needed because low-angle intersections were not present in these networks). In Sect. 3.4, the strain penalty method was tested on Model D, which contains a very low-angle fracture intersection (Figs. 3.27, 3.28, 3.29, and 3.30).

At any location on a (zero-curvature) fracture, strain due to varying displacement discontinuity can be defined as:

$$\varepsilon_k = \frac{dD_k}{dx}, \tag{2.44}$$

where k refers either normal, n, or shear, s, displacement discontinuity and x refers to the distance along the fracture. The derivatives for both modes of deformation are calculated using finite difference approximations at the boundaries between elements. A threshold strain, $\varepsilon_{k,lim}$, is set at each element boundary, and if the absolute value of strain at any element edge exceeds the threshold strain, a penalty stress is applied. The subscript k can refer to n, strain in normal displacement or *s* strain in shear displacement. After a penalty stress is applied, the limit $\varepsilon_{k,lim}$ at the interface is updated to be equal to the absolute value of the current value of ε_k.

The penalty stress is calculated according to:

$$\Delta\sigma_{k,strainadj} = (\varepsilon_k - \varepsilon_{k,\lim})\frac{G}{1 - v}. \tag{2.45}$$

The penalty stress can be applied to both elements at the interface or only to one. The algorithm determines if the motion of each element in the preceding time step acted to increase or decrease the absolute value of the strain at the element interface. If only one element did, the full adjustment is applied to that element. If both did, the adjustment is divided between them proportionally based on their effect on the change in the strain.

The penalty stress is applied at the beginning of the subsequent time step. To prevent excessively large penalty strains from being applied during the subsequent time step, the adaptive time stepping equation, Equation 2.33, is applied with $\delta_{strainadj}$ defined as being equal to the largest absolute value of $\Delta\sigma_{k,strainadj}$, and $\eta_{targ,strainadj}$ defined to be a value equal to one tenth of η_{targ}. As with δ and η_{targ} (explained in Sect. 2.3.9), if the value of $\delta_{strainadj}$ exceeds $4\eta_{targ,strainadj}$, the time step is rejected and repeated with smaller value of *dt*.

2.5.7 Neglecting Stresses Induced by Deformation

A major purpose of the model in this book is to couple deformation with fluid flow. However for purposes of testing and comparison, it is useful to neglect the stresses induced by deformation (Sects. 3.3.1 and 4.2.6). To neglect stresses induced by deformation, all nondiagonal interaction coefficients in the boundary element matrices (Eqs. 2.19, 2.20, 2.21, and 2.22) are set equal to zero. In this case, when an element opens or slides, it affects its own stress, but not the stress at surrounding elements. This assumption has often been made in modeling of shear stimulation, apparently because it simplifies the design of the model significantly (Bruel 1995, 2007; Sausse et al. 2008; Dershowitz et al. 2010).

Neglecting the off-diagonal interaction coefficients causes element stiffness to be discretization dependent because in the Shou and Crouch (1995) method, self-interaction coefficients are a function of element size. To avoid discretization dependence when neglecting stress interaction, self-interaction coefficients are defined so that they are independent of element size. By treating a fracture as a single, constant displacement boundary element, the stiffness of a fracture, K_{frac}, can be calculated as (Crouch and Starfield 1983):

$$K_{frac} = \frac{1}{\pi(1 - v)}\frac{1}{a_{frac}}, \tag{2.46}$$

where a_{frac} is the half-length of the fracture. The self-interaction coefficients of each element for effect of opening deformation on normal stress and shear deformation on shear stress are set to the fracture stiffness given by Eq. 2.46. No adjustment is needed for the self-interaction coefficients that relate opening deformation to shear stress and shear deformation to normal stress because they are always zero.

References

Amestoy, P.R., Davis, T.A., Duff, I.S.: An approximate minimum degree ordering algorithm. SIAM J. Matrix Anal. Appl. **17**(4), 886–905 (1996). doi:10.1137/S0895479894278952

Amestoy, P.R., Davis, T.A., Duff, I.S.: Algorithm 837: AMD, an approximate minimum degree ordering algorithm. ACM Trans. Math. Softw. **30**(3), 381–388 (2004). doi:10.1145/1024074.1024081

Anderson, E., Bai, Z., Bischof, C., Blackford, S., Demmel, J., Dongarra, J., Du Croz, J., Greenbaum, A., Hammarling, S., McKenney, A., Sorensen, D.: LAPACK Users' Guide, 3rd edn. Society for Industrial and Applied Mathematics, Philadelphia (1999)

Aziz, K., Settari, A.: Petroleum Reservoir Simulation. Applied Science Publishers, London (1979)

Barton, N., Bandis, S., Bakhtar, K.: Strength, deformation and conductivity coupling of rock joints. Int. J. Rock Mech. Min. Sci. Geomech. Abstr. **22**(3), 121–140 (1985). doi:10.1016/0148-9062(85)93227-9

Ben-Zion, Y., Rice, J.R.: Earthquake failure sequences along a cellular fault zone in a three-dimensional elastic solid containing asperity and nonasperity regions, J. Geophys. Res. **98**(B8), 14109–14131 (1993), doi:10.1029/93JB01096

Bradley, A.M.: H-matrix and block error tolerances, *arXiv:1110.2807v2*, source code available at http://www.stanford.edu/~ambrad, paper available at http://arvix.org/abs/1110.2807 (2012)

Bruel, D.: Heat extraction modelling from forced fluid flow through stimulated fractured rock masses: application to the Rosemanowes Hot Dry Rock reservoir. Geothermics **24**(3), 361–374 (1995). doi:10.1016/0375-6505(95)00014-H

Bruel, D.: Using the migration of the induced seismicity as a constraint for fractured Hot Dry Rock reservoir modelling. Int. J. Rock Mech. Min. Sci. **44**(8), 1106–1117 (2007). doi:10.1016/j.ijrmms.2007.07.001

Crouch, S.L., Starfield, A.M.: Boundary Element Methods in Solid Mechanics: with Applications in Rock Mechanics and Geological Engineering. Allen & Unwin, London, Boston (1983)

Davis, T.A.: A column pre-ordering strategy for the unsymmetric-pattern multifrontal method. ACM Trans. Math. Softw. **30**(2), 165–195 (2004a). doi:10.1145/992200.992205

Davis, T.A.: Algorithm 832: UMFPACK, an unsymmetric-pattern multifrontal method. ACM Trans. Math. Softw. **30**(2), 196–199 (2004b). doi:10.1145/992200.992206

Davis, T.A.: Direct Methods For Sparse Linear Systems. SIAM, Philadelphia (2006)

Davis, T.A., Duff, I.S.: An unsymmetric-pattern multifrontal method for sparse LU factorization. SIAM J. Matrix Anal. Appl. **18**(1), 140–158 (1997). doi:10.1137/S0895479894246905

Davis, T.A., Duff, I.S.: A combined unifrontal/multifrontal method for unsymmetric sparse matrices. ACM Trans. Math. Softw. **25**(1), 1–19 (1999). doi:10.1145/305658.287640

Dershowitz, W.S., Cottrell, M.G., Lim, D.H., Doe, T.W.: A discrete fracture network approach for evaluation of hydraulic fracture stimulation of naturally fractured reservoirs, ARMA 10-475. Paper presented at the 44th U.S. Rock Mechanics Symposium and 5th U.S.-Canada Rock Mechanics Symposium, Salt Lake City, Utah (2010)

Dieterich, J.H.: Earthquake simulations with time-dependent nucleation and long-range interactions. Nonlinear Process. Geophys. **2**, 109–120 (1995)

Dieterich, J.H., Richards-Dinger, K.B.: Earthquake recurrence in simulated fault systems. In: Savage, M.K., Rhoades, D.A., Smith, E.G.C., Gerstenberger, M.C., Vere-Jones, D. (eds.) Seismogenesis and earthquake forecasting: the frank evison volume II, pp. 233–250, Springer Basel (2010)

Dieterich, J.H.: 4.04—Applications of rate- and state-dependent friction to models of fault slip and earthquake occurrence. In: Gerald, S. (ed.) Treatise on Geophysics pp. 107–129, Elsevier, Amsterdam (2007)

Dongarra, J.J., Croz, J.D., Hammarling, S., Hanson, R.J.: An extended set of FORTRAN Basic Linear Algebra Subprograms. ACM Trans. Math. Softw. **14**(1), 1–17 (1988a). doi:10.1145/42288.42291

Dongarra, J.J., Croz, J.D., Hammarling, S., Hanson, R.J.: Algorithm 656: an extended set of basic linear algebra subprograms: model implementation and test programs. ACM Trans. Math. Softw. **14**(1), 18–32 (1988b). doi:10.1145/42288.42292

Dongarra, J.J., Du Croz, J., Hammarling, S., Duff, I.S.: A set of level 3 basic linear algebra subprograms. ACM Trans. Math. Softw. **16**(1), 1–17 (1990a). doi:10.1145/77626.79170

Dongarra, J.J., Du Croz, J., Hammarling, S., Duff, I.S.: Algorithm 679: a set of level 3 basic linear algebra subprograms: model implementation and test programs. ACM Trans. Math. Softw. **16**(1), 18–28 (1990b). doi:10.1145/77626.77627

Faulkner, D.R., Jackson, C.A.L., Lunn, R.J., Schlische, R.W., Shipton, Z.K., Wibberley, C.A.J., Withjack, M.O.: A review of recent developments concerning the structure, mechanics and fluid flow properties of fault zones. J. Struct. Geol. **32**(11), 1557–1575 (2010). doi:10.1016/j.jsg.2010.06.009

Fredd, C.N., McConnell, S.B., Boney, C.L., England, K.W.: Experimental study of fracture conductivity for water-fracturing and conventional fracturing applications. SPE J. **6**(3), 288–298 (2001). doi:10.2118/74138-PA

Grabowski, J.W., Vinsome, P.K., Lin, R., Behie, G.A., Rubin, B.: A fully implicit general purpose finite-difference thermal model for in situ combustion and steam, SPE 8396. Paper presented at the SPE Annual Technical Conference and Exhibition, Las Vegas, Nevada (1979). doi:10.2118/8396-MS

Holmgren, M.: XSteam: water and steam properties according to IAPWS IF-97. <www.x-eng.com> (2007)

Jaeger, J.C., Cook, N.G.W., Zimmerman, R.W.: Fundamentals of Rock Mechanics, 4th edn. Blackwell Pub, Malden (2007)

Karimi-Fard, M., Durlofsky, L.J., Aziz, K.: An efficient discrete-fracture model applicable for general-purpose reservoir simulators. SPE J. **9**(2), 227–236 (2004). doi:10.2118/88812-PA

Kim, J., Tchelepi, H., Juanes, R.: Stability, accuracy, and efficiency of sequential methods for coupled flow and geomechanics. SPE J. **16**(2), 249–262 (2011). doi:10.2118/119084-PA

Kohl, T., Mégel, T.: Predictive modeling of reservoir response to hydraulic stimulations at the European EGS site Soultz-sous-Forêts. Int. J. Rock Mech. Min. Sci. **44**(8), 1118–1131 (2007). doi:10.1016/j.ijrmms.2007.07.022

Lapusta, N. (2001), Elastodynamic analysis of sliding with rate and state friction, PhD thesis, Harvard University

Lawson, C.L., Hanson, R.J., Kincaid, D., Krogh, F.T.: Basic Linear Algebra Subprograms for FORTRAN usage. ACM Trans. Math. Softw. **5**(3), 308–323 (1979). doi:10.1145/355841.355847

Liu, E.: Effects of fracture aperture and roughness on hydraulic and mechanical properties of rocks: implication of seismic characterization of fractured reservoirs. J. Geophys. Eng. **2**(1), 38–47 (2005). doi:10.1088/1742-2132/2/1/006

McClure, M.W.: Modeling and Characterization of Hydraulic Stimulation and Induced Seismicity in Geothermal and Shale Gas Reservoirs. Stanford University, Stanford, California (2012)

McGarr, A., Simpson, D., Seeber, L.: 40. Case histories of induced and triggered seismicity. In: Lee, W.H.K., Kanamori, H. (eds.) International Geophysics, pp. 647–661, Academic Press (2002)

Morris, J.P., Blair, S.C.: Efficient Displacement Discontinuity Method using fast multipole techniques. Paper presented at the 4th North American Rock Mechanics Symposium, Seattle, WA (2000), <http://www.osti.gov/energycitations/servlets/purl/791449-ZAgLs5/native/>

Olson, J.E.: Predicting fracture swarms—the influence of subcritical crack growth and the crack-tip process zone on joint spacing in rock. Geol. Soc., London, Spec. Publ. **231**(1), 73–88 (2004). doi:10.1144/GSL.SP.2004.231.01.05

Rahman, M.K., Hossain, M.M., Rahman, S.S.: A shear-dilation-based model for evaluation of hydraulically stimulated naturally fractured reservoirs. Int. J. Numer. Anal. Meth. Geomech. **26**(5), 469–497 (2002). doi:10.1002/nag.208

Rice, J.R.: Spatio-temporal complexity of slip on a fault. J. Geophys. Res. **98**(B6), 9885–9907 (1993). doi:10.1029/93JB00191

Rjasanow, S., Steinbach, O.: The Fast Solution of Boundary Integral Equations, 1st edn. Springer, New York (2007)

Sausse, J., Dezayes, C., Genter, A., Bisset, A.: Characterization of fracture connectivity and fluid flow pathways derived from geological interpretation and 3D modelling of the deep seated EGS reservoir of Soultz (France). Paper presented at the Thirty-Third Workshop on Geothermal Reservoir Engineering, Stanford University (2008), <https://pangea.stanford.edu/ERE/db/IGAstandard/record_detail.php?id=5270>

Schultz, R.A.: Stress intensity factors for curved cracks obtained with the Displacement Discontinuity Method. Int. J. Fract. **37**(2), R31–R34 (1988). doi:10.1007/BF00041718

Segall, P.: Earthquake and Volcano Deformation. Princeton University Press, Princeton (2010)

Shou, K.J., Crouch, S.L.: A higher order Displacement Discontinuity Method for analysis of crack problems. Int. J. Rock Mech. Min. Sci. Geomech. Abstr. **32**(1), 49–55 (1995). doi:10.1016/0148-9062(94)00016-V

Teufel, L.W., Clark, J.A.: Hydraulic fracture propagation in layered rock: experimental studies of fracture containment. SPE J. **24**(1), 19–32 (1984). doi:10.2118/9878-PA

Warpinski, N.R., Schmidt, R.A., Northrop, D.A.: In-situ stresses: the predominant influence on hydraulic fracture containment. J. Petrol. Technol. **34**(3), 653–664 (1982). doi:10.2118/8932-PA

Wibberley, C.A.J., Yielding, G., Toro, G.D.: Recent advances in the understanding of fault zone internal structure: a review. Geol. Soc., London, Spec. Publ. **299**(1), 5–33 (2008). doi:10.1144/SP299.2

Willis-Richards, J., Watanabe, K., Takahashi, H.: Progress toward a stochastic rock mechanics model of engineered geothermal systems. J. Geophys. Res. **101**(B8), 17481–17496 (1996). doi:10.1029/96JB00882

Chapter 3
Results

A variety of simulations were performed using four test models: Models A, B, C, and D. The simulations were designed to test the accuracy, convergence, and efficiency of the simulator and to test the effect of a variety of simulation options. In addition, tests were performed to evaluate the accuracy and scaling of Hmmvp for hierarchical matrix decomposition. Section 3.1 gives the details of the simulation settings. Other details of the simulations and results are given in Sects. 3.2, 3.3, 3.4, and 3.5. All simulations were performed using a single processor in a dual quad-core (8 cores) Nehalem CPU running at 2.27 GHz with 24 GB memory as a part of the CEES-Cluster run by the Center for Computational Earth and Environmental Science at Stanford University.

3.1 Simulation and Discretization Details

The discretization settings for all models are given in Table 3.1. Models A1-A4 and D1 used constant element size discretizations, and A5, B, C, and D2 used variable element size discretizations.

Simulations performed with Model A are labeled A[X]-S[Y], where A[X] runs from A1 to A5, and S[Y] runs from S0 to S12. The [X] number refers to the discretization, and the [Y] refers to the particular settings used. Simulations performed with Model B are labeled B1-B9 (all use the same discretization), and the single simulation run with Model C is called C1. Simulations performed with Model D are labeled D[X]-DS[Y], where D[X] could be D1 or D2 (referring to discretization), and DS[Y] could be DS1 or DS2 (referring to settings).

Table 3.2 gives the baseline settings that were the same in all simulations. For each group of simulations (A, B, C, and D), specific baseline settings were defined, given in Table 3.3. Settings for specific Model A and B simulations differed according to settings given in Tables 3.4 and 3.5. Settings for DS1 and DS2 were identical except that DS2 used the strain penalty method.

M. W. McClure and R. N. Horne, *Discrete Fracture Network Modeling*   49
of Hydraulic Stimulation, SpringerBriefs in Earth Sciences,
DOI: 10.1007/978-3-319-00383-2_3, © The Author(s) 2013

Table 3.1 Discretization settings

Model	Total elements	a_{const} (m)	l_c (m)	l_f	l_s	l_o	a_{min} (m)
A1	12	2.5	inf	0	0	0	–
A2	60	0.5	inf	0	0	0	–
A3	300	0.1	inf	0	0	0	–
A4	1,500	0.02	inf	0	0	0	–
A5	48	1	0.3	0.3	0.4	2	0.2
B	52,748	5	0.3	0.3	0.4	2	0.2
C	22,912	5	0.3	0.3	0.4	2	0.2
D1	40	2.5	inf	0	0	0	–
D2	308	2.5	inf	0	0.3	2	0.1

Table 3.2 Baseline settings for all simulations

h	100 m
G	15 GPa
v	0.25
η	3 MPa/(m/s)
μ_f	0.6
$\sigma_{n,Eref}$	20 MPa
$\sigma_{n,eref}$	20 MPa
φ_{Edil}	0°
φ_{edil}	2.5°
$T_{hf,fac}$	10^{-9} m^2
$K_{I,crit}$ (Sect. 2.5.2)	1 MPa$-$m$^{1/2}$
$K_{I,crithf}$ (Sect. 2.5.2)	1 MPa$-$m$^{1/2}$
cstress (Sect. 2.3.3)	Turned off
Crack tip adjustment (Sect. 2.5.2)	Turned on
BEM method (Sect. 2.5.1)	Hmmvp
ε_{tol} (used by Hmmvp, Eq. 3.4)	10^{-6}
Transmissivity updating	Implicit
Friction (Sect. 2.5.3)	Constant (no dynamic weakening)
Adaptive domain adjustment (Sect. 2.5.5)	Not used

3.2 Model A: Small Test Problem

A small test problem (Model A, shown in Fig. 3.1) was used to verify accuracy and convergence and to experiment with different numerical settings. Model A was designed to mimic the opening of splay fractures off a sliding, preexisting flaw (for examples from outcrops, see Segall and Pollard 1983; Mutlu and Pollard 2008).

Constant pressure injection was performed at the center of the middle fracture until the fluid pressure everywhere in the model was equal to the injection pressure. The initial fluid pressure was low enough that none of the fractures were initially open or sliding.

Table 3.3 Model specific baseline settings

	Model A	Model B	Model C	Model D
S_0	0 MPa	0.5 MPa	0.5 MPa	0.5 MPa
$S_{0,\,open}$	0 MPa	0.5 MPa	0.5 MPa	0.5 MPa
E_0	0.1 mm	0.8 mm	0.5 mm	0.5 mm
e_0	0.02 mm	0.03 mm	0.06 mm	0.02 mm
P_{init}	18 MPa	30 MPa	35 MPa	40 MPa
σ_{yy}	26 MPa	75 MPa	75 MPa	55 MPa
σ_{xx}	21 MPa	50 MPa	50 MPa	50 MPa
σ_{xy}	0 MPa	0 MPa	0 MPa	10 MPa
Duration of Simulation	Until P = P_{inj} everywhere	2 h	2 h	Until P = P_{inj} everywhere
P_{injmax}	20.25 MPa	70 MPa	70 MPa	60 MPa
q_{injmax}	None	50 kg/s	50 kg/s	100 kg/s
η_{targ}	0.05 MPa	0.5 MPa	0.5 MPa	0.05 MPa
$D_{e,eff,max}$	10 mm	5 mm	5 mm	5 mm
K_{hf}	Not used	0.01 MPa^{-1}	0.01 MPa^{-1}	Not used
mechtol (Sect. 2.3.5)	0.0003 MPa	0.003 MPa	0.003 MPa	0.003 MPa
itertol (Sect. 2.3.1)	0.001 MPa	0.01 MPa	0.01 MPa	0.01 MPa
Pseudo-3D adjustment (Sect. 2.3.2)	Not used	Used	Used	Used
New fractures (Sect. 2.3.8)	Not permitted	Not permitted	Permitted	Not permitted
Strain Penalty (Sect. 2.5.6)	Not used	Not used	Not used	Used with, $\varepsilon_{n,lim}$ and $\varepsilon_{s,lim} = .001$

Table 3.4 Deviations from baseline settings for Simulations S0-S12

S0	Direct BEM (Hmmvp not used) (Sect. 2.5.1)
S1	$\eta_{targ} = 0.00125\,MPa$ (Sect. 2.3.9)
S2	$\eta_{targ} = 0.0125\,MPa$ (Sect. 2.3.9)
S3	
S4	$\eta_{targ} = 0.5\,MPa$ (Sect. 2.3.9)
S5	$\eta_{targ} = 4\,MPa$ (Sect. 2.3.9)
S6	Direct solution (Sect. 3.2.1), mechtol = 0.0001 (Sect. 2.3.5), itertol = .0002 (Sect. 2.3.1)
S7	Direct solution (Sect. 3.2.1) and uses direct BEM (not Hmmvp), (Sect. 2.5.1), mechtol = 0.0001 (Sect. 2.3.5), itertol = 0.0002 (Sect. 2.3.1)
S8	E_0 = 0.01 mm (Sect. 2.1)
S9	*cstress* option (Sect. 2.3.3) and E_0 = 0.01 mm (Sect. 2.1)
S10	*cstress* option (Sect. 2.3.3) and E_0 = 0.1 mm (Sect. 2.1)
S11	*cstress* option (Sect. 2.3.3) and E_0 = 1 mm (Sect. 2.1)
S12	*cstress* option (Sect. 2.3.3) and E_0 = 10 mm (Sect. 2.1)

Table 3.5 Deviations from baseline settings for Simulations B1-B9

B1	$\eta_{\text{targ}} = 0.05$
B2	$\eta_{\text{targ}} = 0.2$
B3	
B4	$\eta_{\text{targ}} = 4.0$
B5	Adaptive domain adjustment (Sect. 2.5.5)
B6	Dynamic friction weakening with $\mu_d = 0.5$ (Sect. 2.5.3)
B7	No stress transfer (Sect. 2.5.7)
B8	*cstress*, explicit transmissivity updating (Sect. 2.3.3)
B9	Same as B8, also with adaptive domain adjustment (Sects. 2.3.3 and 2.5.5)

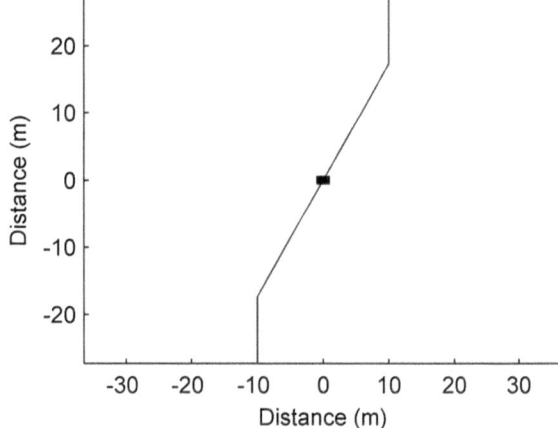

Fig. 3.1 Model A. The *blue lines* represent preexisting fractures. The *black line* represents the wellbore

In order to test the accuracy and convergence of the model's implementation of the Shou and Crouch (1995) Displacement Discontinuity Method, the final displacement distributions were compared to results from an existing code, COMP2DD (Mutlu and Pollard 2008), which implements the Crouch and Starfield (1983) Displacement Discontinuity Method. There was not another code available that could be used for comparison to test the accuracy of the time-dependent part of the calculation. Convergence of the time-dependent results was tested by refining the simulations in time and space and comparing results to the most highly refined solutions.

COMP2DD calculates fracture opening and sliding for given values of remote stress and fluid pressure. Unlike the model described in this book, COMP2DD performs the calculation in a single step; it does not take time steps. In COMP2DD, fluid pressure is assumed to be constant and equal to a value specified by the user. In our model, the fluid pressure changes over time and is variable spatially. To mimic COMP2DD using our model, constant pressure injection was performed until the fluid pressure in every element was the same as the fluid pressure specified in the COMP2DD calculations. Because the final fluid pressures were the same, the final displacements were expected to be the same as long as the results from the model were path-independent.

Results from the model are not necessarily path-independent, but path-dependence does not appear to have played a role in the particular problem used for the code comparison because, as will be shown, the COMP2DD solution was identical to the solution from the model. Stresses caused by opening and sliding are path-independent because deformation is assumed to be elastic. However, determination of the amount of fracture sliding is not path-independent because sliding against friction is not a reversible process: if friction is reduced on a fracture, the fracture will slide, but if friction is subsequently reapplied, the fracture will not slide back to its initial position.

Because elements can be open, sliding, or stationary, COMP2DD does not know ahead of time which boundary conditions to apply (for further discussion, see Sect. 2.3.7). COMP2DD uses a linear complementarity algorithm to handle this issue. Mutlu and Pollard (2008) demonstrated how complementarity has efficiency and accuracy advantages compared to penalty and Lagrange multiplier methods.

In total, 18 simulations were run. Simulations were named according to the both the discretization (A1-A5) and settings (S0-S12): A1-S3, A2-S3, A3-S3, A4-S3, A5-S3, A4-S0, A5-S1, A5-S2, A5-S3, A5-S4, A5-S5, A4-S6, A4-S7, A3-S8, A3-S9, A3-S10, A3-S11, and A3-S12.

The baseline settings used for the simulations are given in Sect. 3.1. Table 3.4 gives specific settings for individual simulations. If not stated explicitly in Table 3.4, the settings used in each simulation were the same as the baseline settings given in Tables 3.2 and 3.3. Models A1-A4 had constant element size with increasing level of refinement. Model A5 had a nonuniform discretization with slightly fewer total elements than A2 (Sect. 3.1).

COMP2DD was used to solve the problem with Model A4, the most finely resolved discretization (Sect. 3.1). The constant displacement method used by COMP2DD (Crouch and Starfield 1983) is lower order than the quadratic method used in the simulator (Shou and Crouch 1995), and so for the same discretization, COMP2DD should be less accurate. However, the accuracy of the simulator is limited by the convergence criteria used in the iterative solvers (Sects. 2.3.1, 2.3.4, 2.3.5). Therefore, the COMP2DD solution should not be considered an "exact" solution because its results may be of comparable order of accuracy as the most precise results from the model. To demonstrate the accuracy of the model, it is sufficient to show that with refinement the results converge to an answer that is very close to the COMP2DD result.

The COMP2DD result was compared to results from Simulations A1-S3, A2-S3, A3-S3, A4-S3, and A5-S3. Figures 3.2 and 3.3 show the final opening and sliding distributions from the simulations and COMP2DD.

The relative difference, e_j, between the results from COMP2DD and from the model was quantified by comparing the final values of opening and sliding at different points along the fracture. The comparison points were at the center of each element in the discretization. When comparing results, the term "relative difference" is used instead of "error" because the COMP2DD solution is not an exact solution. Because the A5 discretization was nonuniform, the element centers

Fig. 3.2 Final sliding
distribution along a section of
Model A for COMP2DD
(*black*), A1-S3 (*blue*), A2-S3
(*green*), A3-S3 (*cyan*), A4-S3
(*maroon*), and A5-S3 (*red*).
The COMP2DD and A4-S3
lines coincide

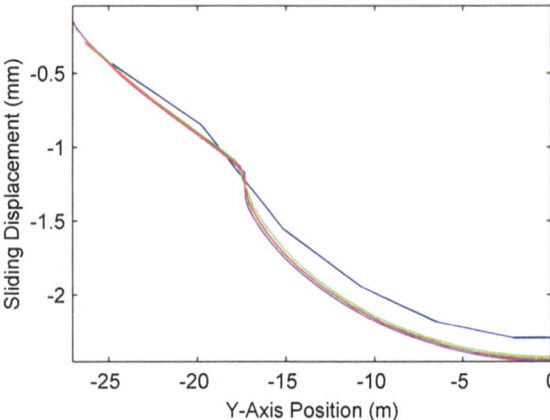

Fig. 3.3 Final opening
distribution along a section of
Model A for COMP2DD
(*black*), A1-S3 (*blue*), A2-S3
(*green*), A3-S3 (*cyan*), A4-S3
(*maroon*), and A5-S3 (*red*).
The COMP2DD and A4-S3
lines coincide

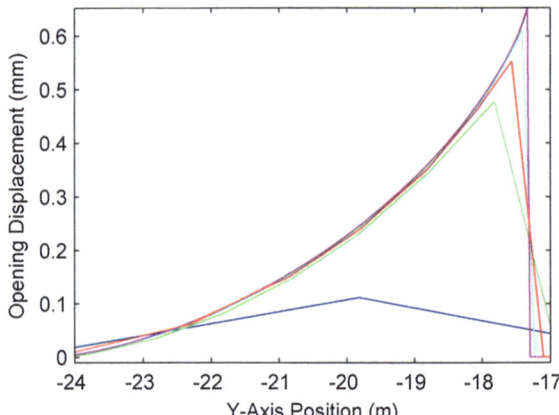

of the COMP2DD solution (which used the A4 discretization) and the A5 discretization did not coincide. For comparison to the A5 solutions, COMP2DD values were linearly interpolated to the A4 grid. Relative difference was calculated according to the expression:

$$e_j = \sqrt{\sum_i^{N_j} [(((D_i^0 - D_i^j)/|D^0|_{max})^2 + ((E_i^0 - E_i^j)/|E^0|_{max})^2)/(2N_j)]} \quad (3.1)$$

where j refers to the particular result being compared (A1-S1 to A5-S1), i refers to a particular nodal point on the discretization, the superscript ^{0} refers to the COMP2DD solution (interpolated to the grid of j in the case of A5-S1), N_j refers to the total number of elements in grid j, D refers to shear displacement, and E refers to opening displacement. The values $|D^0|_{max}$ and $|E^0|_{max}$ refer to the largest absolute values of D and E in the COMP2DD solution, about 0.65 mm for opening and about 2.45 mm for sliding, and are used to normalize the results.

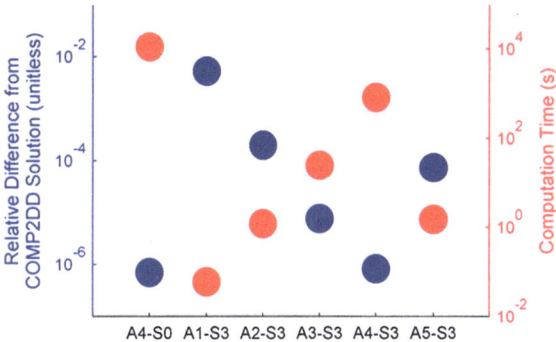

Fig. 3.4 Computation time (*red*) and relative difference (*blue*) from COMP2DD for Simulations A4-S0, A1-S3, A2-S3, A3-S3, A4-S3, and A5-S3. Relative difference was calculated according to Eq. 3.1 on the basis of final fracture deformation and normalized by reference values

The relative differences and computation times for A4-S0, A1-S3, A2-S3, A3-S3, A4-S3, and A5-S3 are shown in Fig. 3.4.

To identify the effect of spatial discretization on accuracy of the time-dependent results, flow rate over time was compared for different spatial discretizations (using the same temporal discretization factor η_{targ}). Figure 3.5 is a plot of injection rate versus time for Simulations A1-S3, A2-S3, A3-S3, A4-S3, and A5-S3.

The relative difference for different discretizations was calculated by comparing the various solutions to the most finely refined solution, A4-S3. Values of flow rate were interpolated onto a temporal discretization spaced one second apart for the first 5000 s of the simulation. Relative difference in flow rate was calculated according to:

$$e_j = \sqrt{\frac{1}{N} \sum_i^N (Q_i^0 - Q_i^j)^2},\tag{3.2}$$

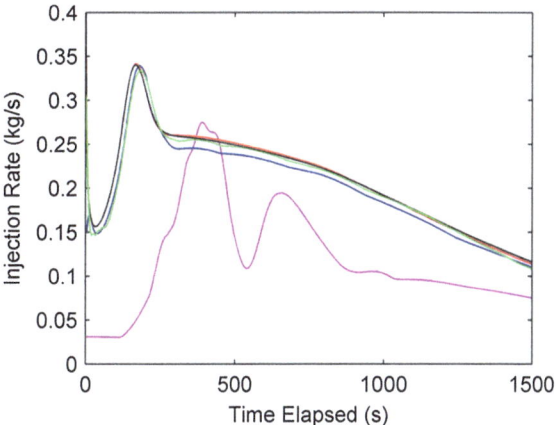

Fig. 3.5 Injection rate versus time for Simulations A1-S3 (*maroon*), A2-S3 (*blue*), A3-S3 (*red*), A4-S3 (*black*), and A5-S3 (*green*)

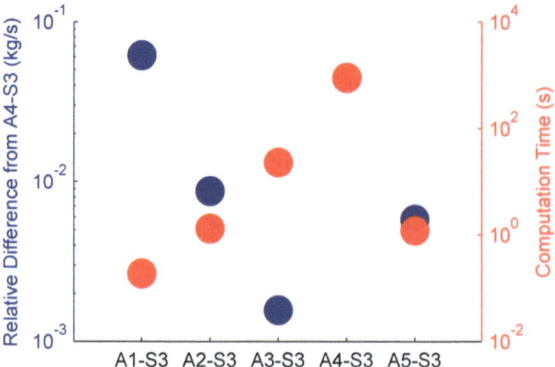

Fig. 3.6 Computation time (*red*) and relative difference (*blue*) in flow rate versus time for Simulations A1-S3, A2-S3, A3-S3, A4-S3, and A5-S3. Relative difference calculated according to Eq. 3.2 on the basis of injection rate versus time

where the superscript ^{0} refers to Simulation A4-S3, Q is the injection rate, and N refers to the 5000 points in time where Q was sampled. Relative difference and computation time are shown in Fig. 3.6.

To isolate the effect of temporal discretization, simulation settings S1-S5 were used with Model A5. The simulations were identical except that they used different values of η_{targ}. Relative difference was calculated by comparing the flow rate over time between the various simulations and Simulation A5-S1, which used the smallest η_{targ}. Because Model A5 was only moderately well refined spatially, all solutions had some error, regardless of temporal discretization. Figure 3.7 shows plots of injection rate versus time for the various simulations. Figure 3.8 shows the relative differences (compared to A5-S1) and the computation times for the various simulations.

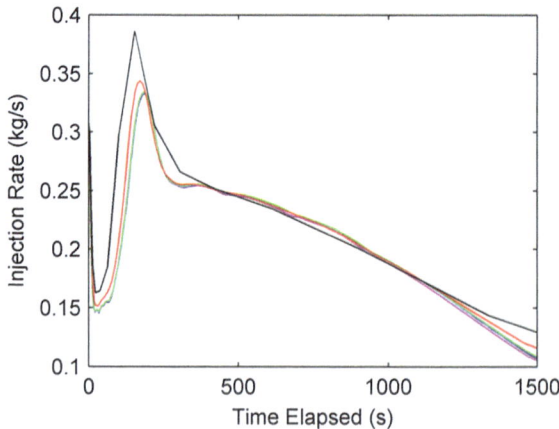

Fig. 3.7 Injection rate versus time for A5-S1 (*maroon*), A5-S2 (*blue*), A5-S3 (*green*), A5-S4 (*red*), and A5-S5 (*black*)

3.2.1 Solving Directly for the Final Deformations

It is possible to modify the model to directly solve for the final fracture displacements in a single step (what COMP2DD does) instead of using time steps. This approach could be useful if the objective of the calculation is to determine the final displacement field, and the intermediate displacements are not of interest. Because the model was not designed to be used in this way, it is necessary to use several unusual settings. In order to maintain constant fluid pressure as the fractures deform, the fluid compressibility is made extremely large (effectively infinite) and the transmissivity is set to zero. The fluid flow and normal stress subloop convergence criteria are modified to depend only on the residual of the stress equations and not the mass balance equations. Finally, the radiation damping coefficient, η, is set to zero. With these settings, the problem can be solved in a single "time step" (the duration of the time step has no effect on the solution), and yields the same solution as COMP2DD.

Two simulations using discretization A4 were run with the direct solution method, S6 and S7. S6 used Hmmvp for approximate matrix multiplication, and S7 used the direct BEM (see Sect. 2.5.1). A comparison of the results is given in Fig. 3.9. Two different complementarity algorithms implemented in COMP2DD were used, Lemke (Ravindran 1972) and SOCCP (Hayashi et al. 2005).

3.2.2 Testing the Effect of cstress

Simulations A3-S9, A3-S10, A3-S11, and A3-S12 were performed with the *cstress* option (Sect. 2.3.3). In S9-S12, different values for E_0 were used. The simulations using S9-S12 can be compared to Simulation A3-S8, which was identical to A3-S9 but did not use the *cstress* option.

With the *cstress* option activated, stresses induced by the normal deformation of closed fractures are not neglected. Fractures with larger values of E_0 contain a

Fig. 3.8 Computation time (*red*) and relative difference (*blue*) from A5-S1 for Simulations A5-(S1-S5). Relative difference calculated according to Eq. 3.2 on the basis of injection rate versus time

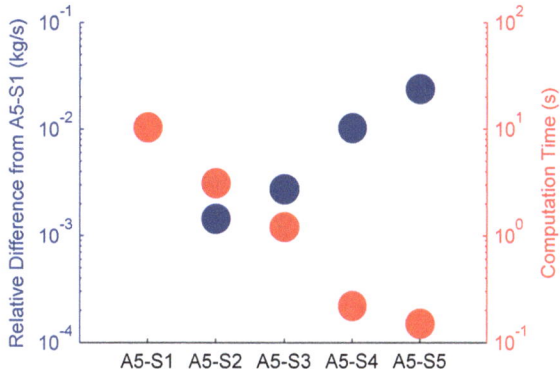

Fig. 3.9 Computation time
(*red*) and relative difference
(*blue*) from A4-SOCCP for
A4-S6, A4-S7, and A4-
LEMKE, and computation
time for A4-SOCCP. Relative
difference calculated
according to Eq. 3.1 on the
basis of final fracture
deformation and normalized
by reference values

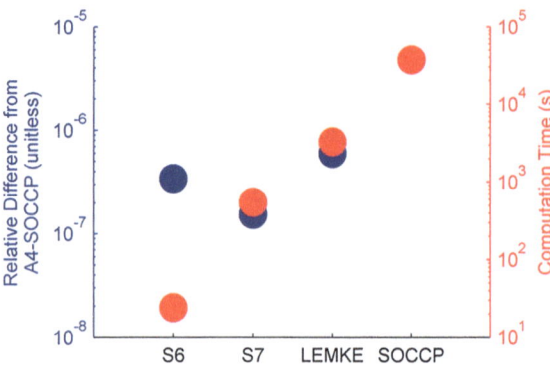

Fig. 3.10 Final opening
distribution along a section of
Model A for A3-S8 (*blue*),
A3-S9 (*green*), A3-S10 (*red*),
A3-S11 (*black*), and A3-S12
(*maroon*). Models A3-S8 and
A3-S9 coincide

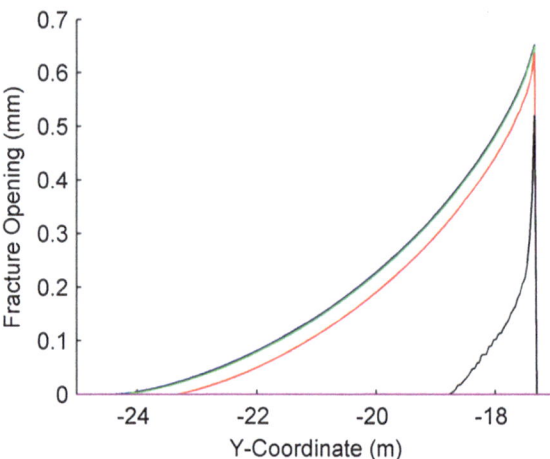

greater volume of fluid and experience greater normal deformation (and induce
greater stresses) for the same perturbation in fluid pressure. The final displacements of the simulations are shown in Figs. 3.10 and 3.11.

3.3 Models B and C: Large Test Problems

Several simulations were performed on large fracture networks, Models B and C.
Model B was designed to demonstrate shear stimulation. In shear stimulation,
injection enhances transmissivity by inducing slip on preexisting fractures (Pine
and Batchelor 1984; Evans 2005; Cladouhos et al. 2011; Sect. 3.1.1 of McClure
2012). Model C was designed to demonstrate mixed mode propagation—where
injection causes shear and opening of preexisting fractures as well as propagation
of new opening mode fractures (Sect. 3.1.1 of McClure 2012). The discretization

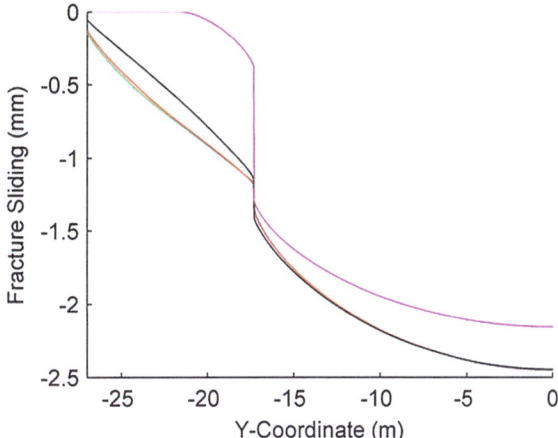

Fig. 3.11 Final sliding distribution along a section of Model A for A3-S8 (*blue*), A3-S9 (*green*), A3-S10 (*red*), A3-S11 (*black*), and A3-S12 (*maroon*). Models A3-S8 and A3-S9 coincide

settings used for Models B and C are given in Sect. 3.1. Models B and C are shown in Figs. 1.1 and 2.4.

In the Model B and C simulations, injection was carried out for two hours at a constant injection rate. The simulation was stopped at the end of two hours. The post-injection redistribution of fluid pressure was not included in the simulation.

During the Model B simulations, the fluid pressure exceeded the least principal stress, a condition that should have led to the propagation of new tensile fractures. However, "potentially forming fractures" were not specified in the fracture network, and so new fractures were not able to form. This unrealistic model behavior is acceptable for the purposes of this book, because the simulations in this book are intended only to test the accuracy and convergence of the numerical simulator. Even though there was not any propagation of new fractures, opening was able to occur on preexisting fractures.

3.3.1 Model B: Large Test Problem of Shear Stimulation

Nine simulations were performed with Model B, B1-B9, and one simulation, Simulation C1, was performed with Model C. The objective of these simulations was to test the effect of simulation options on the results and efficiency of the model.

The baseline settings for all simulations are given in Table 3.3. The baseline settings for all Model B simulations are given in Table 3.4. The deviations from baseline for each individual simulation B1-B9 are given in Table 3.5.

The objective of Simulations B1-B4 was to test the efficiency and accuracy of simulations with various values of η_{targ}. This parameter affects how many time steps are taken during the simulation (Sect. 2.3.9). Figures 3.12 and 3.13 show the final shear displacement (proportional to color) and opening (proportional to

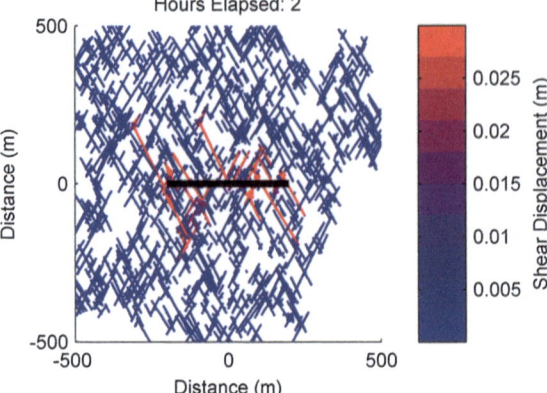

Fig. 3.12 Final fracture shear displacement and opening (*thickness* is proportional to opening) of Simulation B1, the most highly temporally-resolved simulation

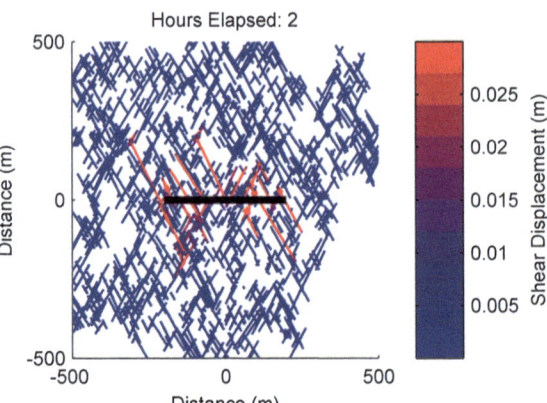

Fig. 3.13 Final fracture shear displacement and opening (*thickness* is proportional to opening) of Simulation B4, the most poorly temporally-resolved simulation

thickness, but exaggerated scale) for Simulations B1 and B4, the most and least temporally resolved simulations. The fracture opening and sliding distributions for B2 and B3 are not shown because they appear visually identical to Fig. 3.12.

The relative difference in sliding displacement between Models B1 and Models B2-B5, B8 and B9 is shown in Fig. 3.14. Model B5 used the same value of η_{targ} as B3 but used adaptive domain adjustment (Sect. 2.5.5). Models B8 and B9 used *cstress* (Sect. 2.3.3), and Model B9 also used adaptive domain adjustment. Relative difference in sliding displacement was calculated according to:

$$e_j = \sqrt{\frac{1}{M} \sum_i^M ((D_i^0 - D_i^j))^2},\qquad(3.3)$$

where D refers to the sliding displacement, ^{0} refers to the result from Simulation B1, i refers to a particular element, j refers to the particular simulation, $|D^0|_{avg}$ is a scaling displacement, and M refers to the number of comparison points. The

Fig. 3.14 Relative difference between sliding displacement in Simulations B2-B5, B8 and B9 compared to Simulation B1. Relative difference calculated according to Eq. 3.3

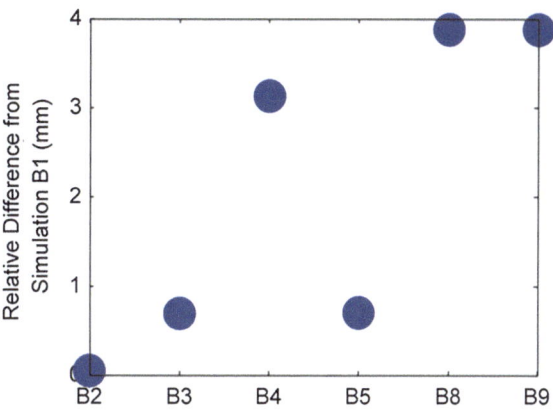

comparison points were the elements from Simulation B1 that had a sliding displacement with magnitude greater than 1.0 mm. The average absolute value of the displacements at the comparison points was 2.27 cm.

Model B6 used dynamic friction weakening (Sect. 2.5.3). Model B7 neglected all stress transfer—allowing elements to deform in response to changes in fluid pressure, but not updating stresses on surrounding elements due to those deformations (Sect. 2.5.7). Model B8 used the *cstress* option (Sect. 2.3.3). Model B9 used the *cstress* option and adaptive domain adjustment (Sects. 2.3.3 and 2.5.5).

The final sliding and opening distributions of Models B6-B8 are shown in Figs. 3.15, 3.16, and 3.17. The sliding and opening distribution for Model B9 is not shown because it appears visually identical to the sliding and opening distribution for Model B8, shown in Fig. 3.17.

In Simulation B6, dynamic friction weakening caused seismic events to occur. By definition, a seismic event occurred whenever the sliding velocity of the fastest sliding element in the model was above a certain threshold, 5 mm/s. When the

Fig. 3.15 Final fracture shear displacement and opening (*thickness* is proportional to opening) of Simulation B6, which used dynamic friction weakening

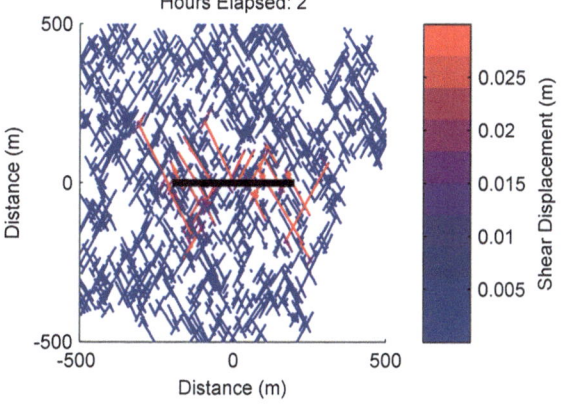

Fig. 3.16 Final fracture shear displacement and opening (*thickness* is proportional to opening) of Simulation B7, which neglected stresses induced by deformation of fracture elements

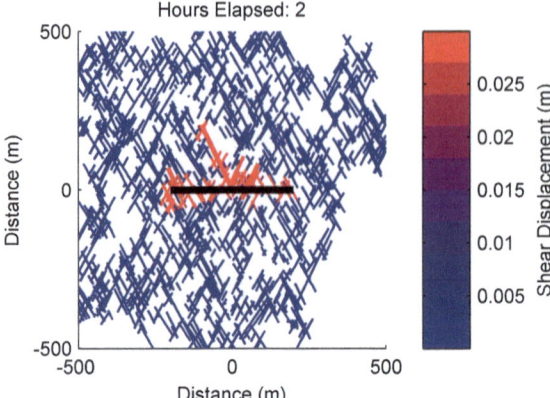

Fig. 3.17 Final fracture shear displacement and opening (*thickness* is proportional to opening) of Simulation B8, which used the *cstress* option

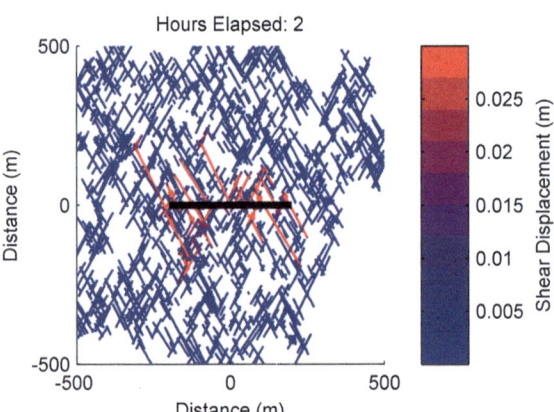

fastest sliding element in the model had a velocity below 5 mm/s, by definition, the seismic event ended. The simulator tracked the total cumulative slip that took place at velocity greater than 5 mm/s during each seismic event, D_{cum}. After the seismic event ended, the seismic moment of the event, M_0, was calculated (Hanks and Kanamori 1979):

$$M_0 = G \int D_{cum} dA, \qquad (3.4)$$

where A is fracture surface area. The moment was correlated to a moment magnitude, M_w, according to Hanks and Kanamori (1979):

$$M_w = \frac{\log_{10} M_0}{1.5} - 6.06, \qquad (3.5)$$

where M_0 is expressed in Newton-meters.

Fig. 3.18 Seismic event
magnitude versus time for
Simulation B6. Magnitude
calculated according to
Eq. 3.5

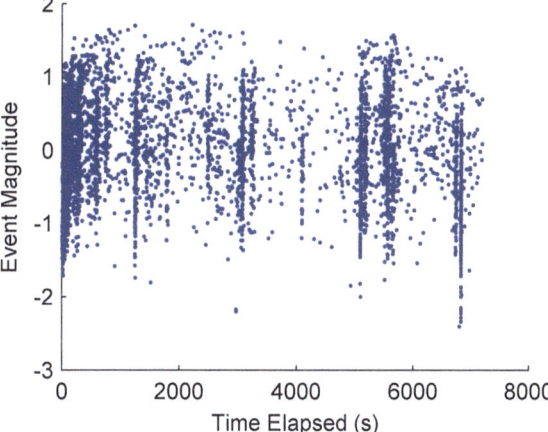

Fig. 3.19 Magnitude-
frequency distribution during
Simulation B6

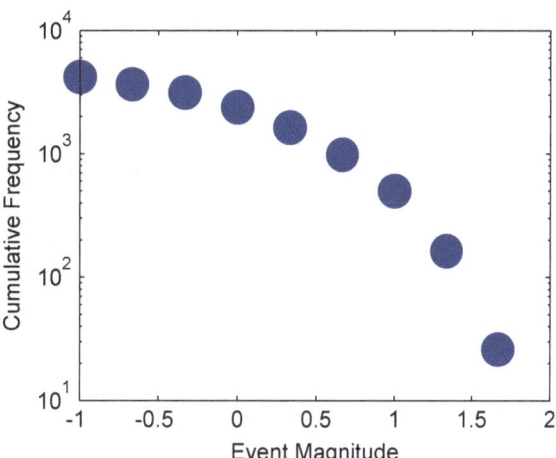

Typical seismic event durations were 0.001–0.01 s for the smaller events and as
long as 0.2 s for the largest events. Figure 3.18 shows a plot of event magnitude
versus time for Simulation B6.

Figure 3.19 shows the magnitude-frequency distribution during Simulation B6.
The trend in Fig. 3.19 is curving, and so we conclude that it does not replicate a
Gutenberg-Richter magnitude frequency distribution. Parameters could probably
have been tuned to force the magnitude-frequency distribution to be linear, but this
would have been a model overfit, not a meaningful model result.

Seismic events were usually confined to a single fracture, and so the magnitude-
frequency distribution for larger events was controlled by the fracture size distribution. The smallest events occurred on a single element, and so the magnitude-

Fig. 3.20 Seismic event
hypocenters (with adjustment
to approximate the effect of
relocation error) for
Simulation B6. Larger
symbols represent larger
magnitude events

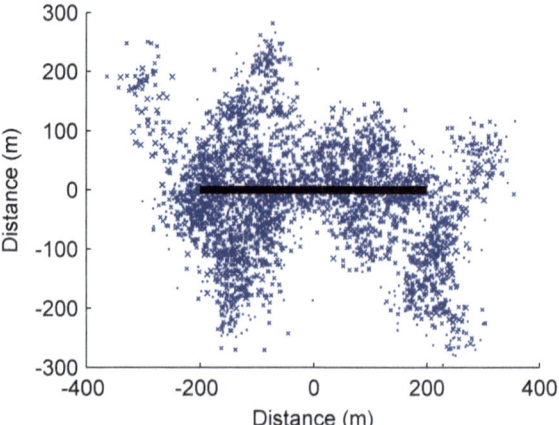

Fig. 3.21 Computer run-
time (*blue*) and total number
of time steps (*red*) for
Simulations B1-B9

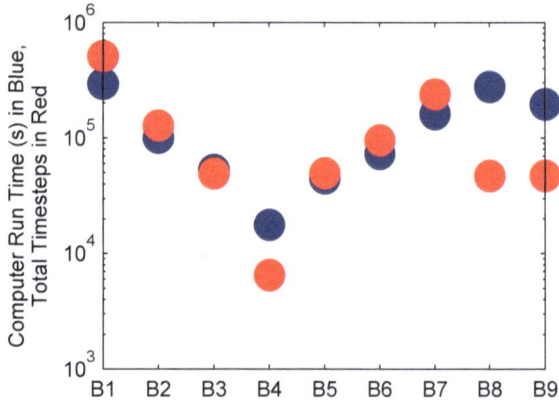

frequency distribution for small events was controlled by the distribution of element sizes. The magnitude-frequency distribution in the intermediate range was
affected by the tendency for slip on a small number of fractures to cascade into
larger slip events and the tendency for larger cascades to stop propagating.

Figure 3.20 shows the location of the event hypocenters during Simulation B6.
To approximate the effect of relocation error, the hypocenters were relocated
randomly within a circle with radius of 30 m from the actual hypocenter. For each
seismic event, the hypocenter was defined as the first location where sliding
velocity increased above 5 mm/s.

Figure 3.21 shows the total computer run-time and total number of time steps
taken during Simulations B1-B9. Figure 3.22 shows the average computer run-
time per time step taken during Simulations B1-B9.

Figure 3.23 shows a plot of total time steps performed versus computer run-
time for Simulations B3, B5, B8, and B9. Simulations B8 and B9 used the *cstress*

Fig. 3.22 Average computer run-time per time step for Simulations B1-B9

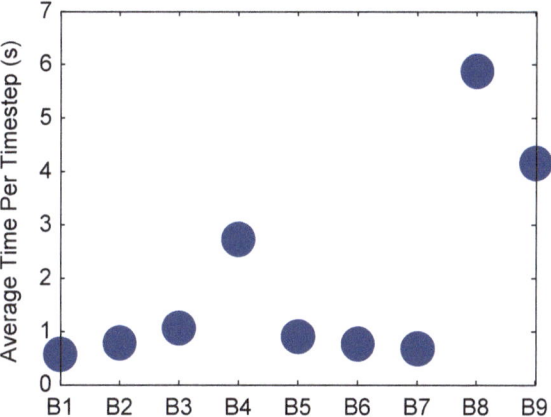

Fig. 3.23 Total number of time steps performed versus run-time for Simulations B3 (*blue*), B5 (*red*), B8 (*black*), and B9 (*green*)

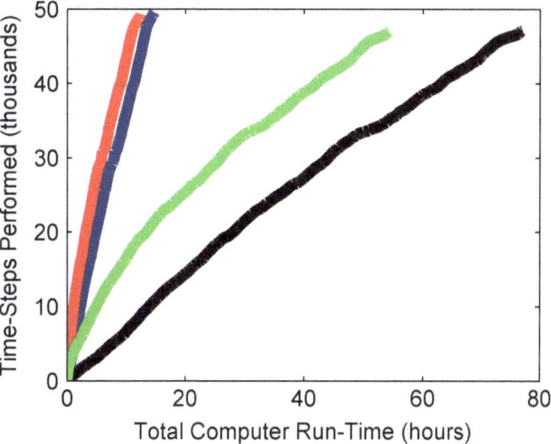

option, and B3 and B5 did not. B3 and B5 are identical except that the latter used adaptive domain adjustment. B8 and B9 are identical except that B9 used adaptive domain adjustment.

Figure 3.24 shows injection pressure versus time for Simulations B1-B4. Figure 3.25 shows injection pressure versus time for Simulations B6-B8. In all simulations, injection pressure was constant at 50 kg/s for the entire simulations (except a brief period at the initiation of injection).

Fig. 3.24 Injection pressure versus time for Simulations B1 (*blue*), B2 (*red*), B3 (*green*), and B4 (*black*)

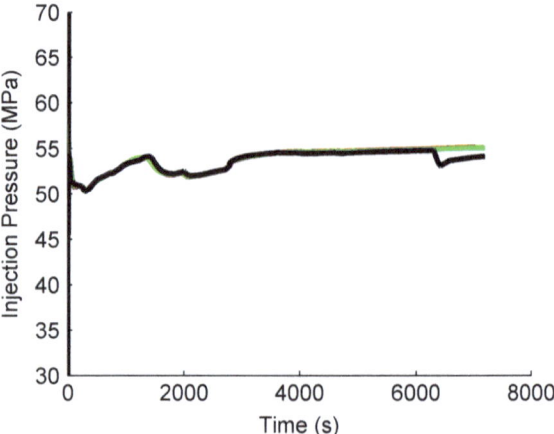

Fig. 3.25 Injection pressure versus time for Simulations B6 (*blue*), B7 (*red*), and B8 (*green*)

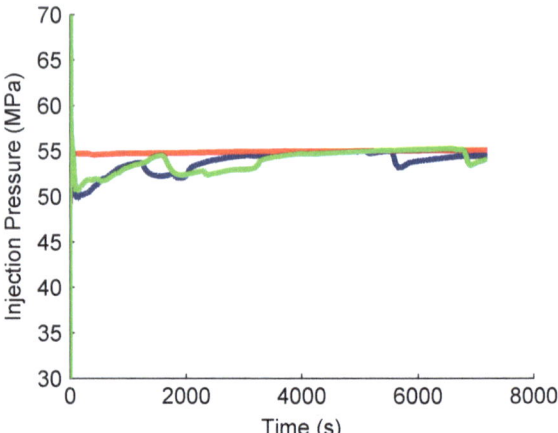

3.3.2 Model C: Large Test Problem of Mixed-Mode Stimulation

In Simulation C1, propagation of new opening mode fractures occurred (Sect. 2.3.8), in contrast to the simulations using Models A, B, and D, in which no new fractures formed. Figure 2.4 shows a realization of both the potentially forming and the preexisting fractures in Model C. Settings for Simulation C1 are given in Sect. 3.1. Figure 3.26 shows the final transmissivity and opening distribution for Simulation C1. The computer run-time was 15,992 s (about four hours) and 15,574 time steps were taken.

Fig. 3.26 Final
transmissivity and opening
distribution (*thickness* is
proportional to opening, but
not to scale) for Simulation
C1. The Wellbore (not
shown) was located from
$(-200,0)$ to $(200,0)$

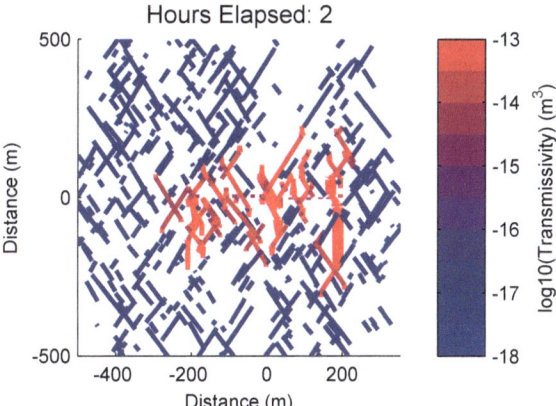

3.4   Model D: Testing the Strain Penalty Method

Four simulations using Model D were performed to test the strain penalty method
described in Sect. 2.5.6. Two discretizations were used, D1 and D2, (described in
Sect. 3.1), and two settings files were used, DS1 and DS2. The settings used in
DS1 and DS2 are given in Sect. 3.1. DS1 and DS2 were identical except that the
strain penalty method was used in the latter and not the former. Discretization D1
was coarsely refined. D2 was significantly refined, especially around the
intersection.

Figures 3.27, 3.28, 3.29, and 3.30 shows the final shear displacement and
opening distributions for the four simulations, D1-DS1, D1-DS2, D2-DS1, and D2-
DS2.

Fig. 3.27 Final shear
displacement and opening
displacement (*thickness* is
proportional to opening, but
not to scale) for Simulation
D1-DS1. Note that the *x-axis*
scale and the *y-axis* scale are
different

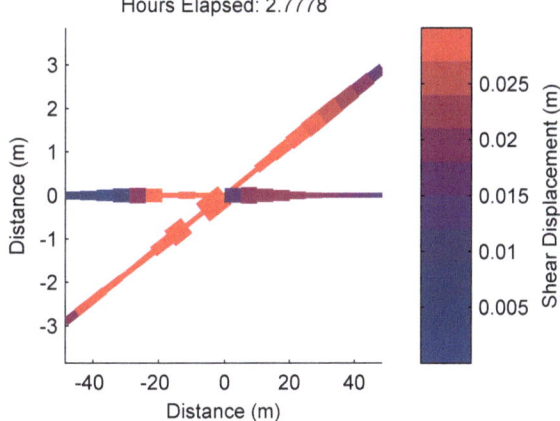

Fig. 3.28 Final shear
displacement and opening
displacement (*thickness* is
proportional to opening, but
not to scale) for Simulation
D1-DS2. Note that the *x-axis*
scale and the *y-axis* scale are
different

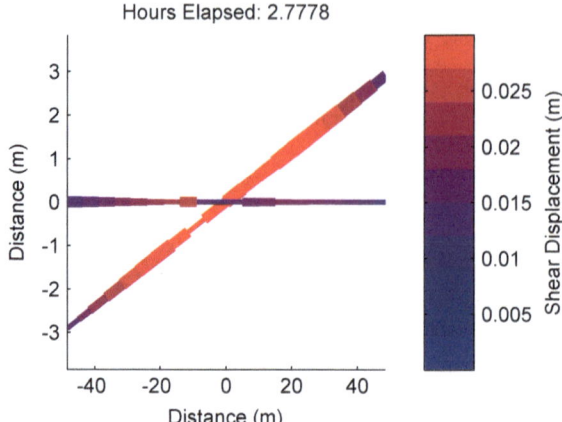

Fig. 3.29 Final shear
displacement and opening
displacement (*thickness* is
proportional to opening, but
not to scale) for Simulation
D2-DS1. Note that the *x-axis*
scale and the *y-axis* scale are
different

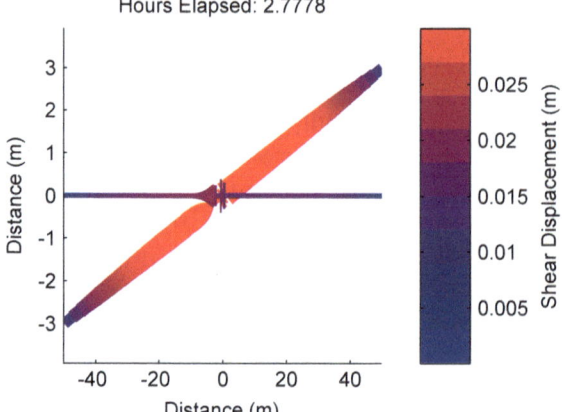

Fig. 3.30 Final shear
displacement and opening
displacement (*thickness* is
proportional to opening, but
not to scale) for Simulation
D2-DS2. Note that the *x-axis*
scale and the *y-axis* scale are
different

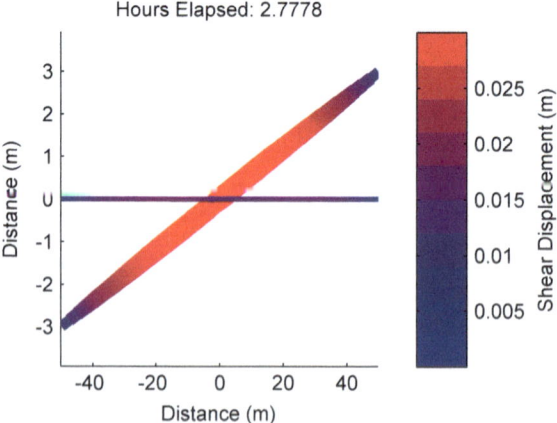

3.5 Hierarchical Matrix Decomposition

A variety of tests were performed to measure the accuracy and compression Hmmvp, the BEM approximation code Bradley (2012). Plots of floating point operations (FLOPs) per matrix multiplication versus problem size were created for two cases: increasing discretization refinement for a given fracture network and increasing fracture network size for a given degree of discretization refinement. FLOPs per matrix multiplication is defined as being the number of additions and multiplications required to update both the normal and shear stress of all elements in a model due to normal and shear displacements of all elements. For n elements, a direct matrix multiplication would require $8n^2$ FLOPs. FLOPs per matrix multiplication is a close proxy for the memory required to store the interaction coefficients of the matrix decompositions.

In Hmmvp, a relative tolerance ε_{tol} of 10^{-6} was specified, such that:

$$\frac{||B\Delta D - B_h \Delta D||_2}{||B||_F ||\Delta D||_2} \leq \varepsilon_{tol}, \qquad (3.6)$$

where B represents one of the full matrices of interaction coefficients, and B_h represents the Hmmvp matrix approximation Bradley (2012). The subscript F reflects the Frobenius norm of a matrix and the subscript 2 reflects the Euclidian norm of a vector. Frobenius and Euclidian norms are defined as the square root of the sum of squares of all values in a matrix or vector, respectively. Testing on a variety of fracture networks with randomly generated displacement vectors demonstrated that the relative error, e_{hmat}, never exceeded 0.001, where:

$$e_{hmat} = \frac{||B\Delta D - B_h \Delta D||_2}{||B\Delta D||_2}. \qquad (3.7)$$

To test increasing refinement on a given fracture network, the fracture network shown in Fig. 3.31 was used. The network contains 237 fractures. A variety of discretizations were created using different values for l_s, l_o, l_f, l_c, and minimum element size. Table 3.6 gives the settings used for each discretization and the total number of elements in each discretization. Each discretization was decomposed using Hmmvp. Figure 3.32 shows the number of FLOPS required to perform a matrix multiplication with Hmmvp method and with direct multiplication.

Next, a scaling comparison was performed for different sized fracture networks using the same level of discretization. A large network was generated with 6000 fractures and then different subsets of the network were used. Figure 3.33 shows the full fracture network with black rectangles showing the subdivisions of the network that were used for comparison. The discretizations were created using $l_s = 0.4$, $l_o = 4.0$, $l_f = 0.7$, $l_c = 0.5$ m, and a_{min} equal to 0.1 m. Figure 3.34 shows the number of FLOPS for a matrix multiplication using each discretization.

The amount of time required to generate the matrix decompositions does not affect the efficiency of the simulator because the decompositions are performed

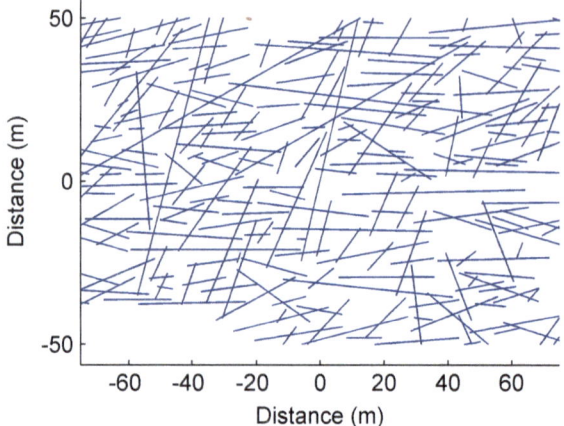

Fig. 3.31 Fracture network used for discretization comparison at different levels of refinement

Table 3.6 Settings used for the discretizations in Fig. 3.32

N	l_s	l_o	l_c (m)	l_f	a_{min}
872	0	0	inf	0.7	0.4
3966	0	0.5	0.5	0.7	0.4
4646	0.1	2	0.5	0.7	0.4
6352	0.2	2	0.5	0.7	0.4
15872	0.4	4	0.5	0.7	0.1
64192	0.6	4	0.5	0.7	0.05

Fig. 3.32 Comparison of FLOPS/multiplication for different levels of refinement on the fracture network shown in Fig. 3.31. Shown in *blue* is the full matrix multiplication. The Hmmvp result is shown in *red*. The *green line* is linear with a slope of one and is shown for reference

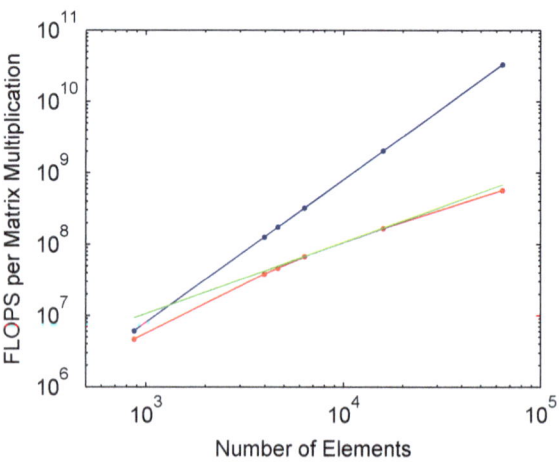

Fig. 3.33 Fracture network used the test matrix approximation scaling for a variety of network sizes. *Black boxes* show the boundaries of each fracture network

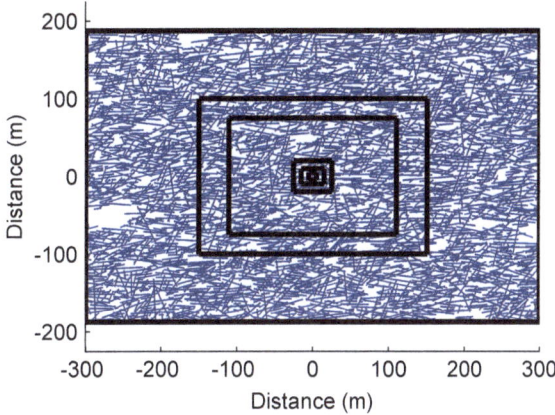

Fig. 3.34 Comparison of FLOPS/multiplication for the same discretization refinement on different sized fracture networks shown in Fig. 3.33. The full matrix multiplication is shown in *blue*. The Hmmvp result is shown in *red*. The *green line* scales like $n\log(n)$ and is shown for reference

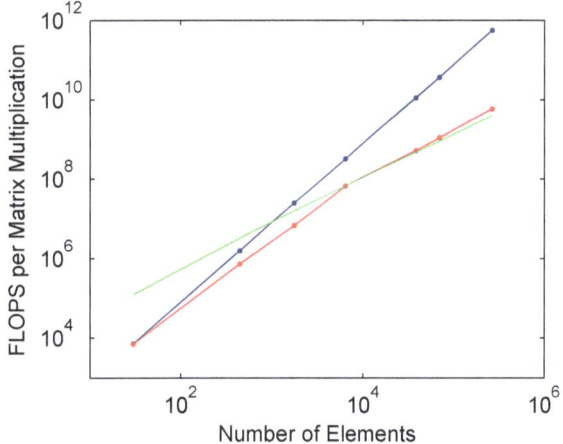

Fig. 3.35 Amount of time required by Hmmvp to calculate the matrix decompositions. The *red line* shows the networks of variable sizes shown in Fig. 3.33. The *blue line* shows the results for the variable refinements shown in Fig. 3.31. The *green line* is linear with a slope of one and is shown for reference

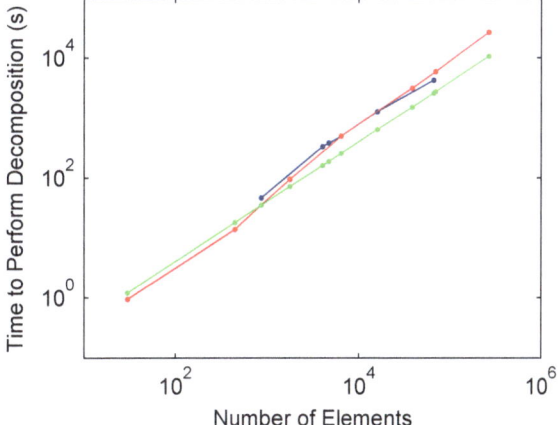

once, prior to simulation. However, the decompositions must be efficient enough that they can be performed in a practical duration of time. Figure 3.35 shows the amount of time required to perform the decompositions discussed in this section.

References

Bradley, A.M.: H-matrix and block error tolerances, arXiv:1110.2807v2, (2012). source code available at http://www.stanford.edu/~ambrad, paper available at http://arvix.org/abs/1110.2807

Cladouhos, T.T., Clyne, M., Nichols, M., Petty, S., Osborn, W.L., Nofziger, L.: Newberry Volcano EGS demonstration stimulation modeling. Geoth. Res. Counc. Trans. **35**, 317–322 (2011)

Crouch, S.L., Starfield, A.M.: Boundary element methods in solid mechanics: with applications in rock mechanics and geological engineering, Allen and Unwin, London Boston (1983)

Evans, K.F.: Permeability creation and damage due to massive fluid injections into granite at 3.5 km at Soultz: 2. Critical stress and fracture strength, J. Geophys. Res. **110**(B4) (2005). doi:10.1029/2004JB003169

Hanks, T.C., Kanamori, H.: A moment magnitude scale, J. Geophys. Res. **84**(B5), 2348–2350 (1979). doi:10.1029/JB084iB05p02348

Hayashi, S., Yamashita, N., Fukushima, M.: A combined smoothing and regularization method for monotone second-order cone complementarity problems. SIAM J. Optim. **15**(2), 593–615 (2005). doi:10.1137/S1052623403421516

McClure, M.W.: Modeling and characterization of hydraulic stimulation and induced seismicity in geothermal and shale gas reservoirs. Stanford University, Stanford, California (2012)

Mutlu, O., Pollard, D.D.: On the patterns of wing cracks along an outcrop scale flaw: a numerical modeling approach using complementarity, J. Geophys. Res. **113**(B6) (2008). doi:10.1029/2007JB005284

Pine, R.J., Batchelor, A.S.: Downward migration of shearing in jointed rock during hydraulic injections. Int. J. Rock Mech. Min. Sci. Geomech. Abstr. **21**(5), 249–263 (1984). doi:10.1016/0148-9062(84)92681-0

Ravindran, A.: Algorithm 431: a computer routine for quadratic and linear programming problems [H]. Commun. ACM **15**(9), 818–820 (1972). doi:10.1145/361573.1015087

Segall, P., Pollard, D.D.:Nucleation and growth of strike slip faults in granite, J. Geophys. Res. **88**(B1), 555–568 (1983). doi:10.1029/JB088iB01p00555

Shou, K.J., Crouch, S.L.: A higher order Displacement Discontinuity Method for analysis of crack problems. Int. J. Rock Mech. Min. Sci. Geomech. Abstr. **32**(1), 49–55 (1995). doi:10.1016/0148-9062(94)00016-V

Chapter 4
Discussion

4.1 Model A

4.1.1 General Description of Results

In the Model A simulations, injection was performed at constant pressure (less than the least principal stress) into the center of the fracture shown in Fig. 3.1. The simulations ended when the entire system reached the injection pressure. As in all simulations, matrix permeability was assumed negligible, so fluid did not leak off from the fractures.

Injection reduced the effective normal stress on the fractures, inducing slip on the central fracture. The neighboring fractures did not initially bear shear stress because they were perpendicular to the least principal stress, but the slip of the central fracture induced shear stress on the neighboring fractures and caused them to slip (Fig. 3.2). Slip of the central fracture also induced tensile stress on the neighboring fractures and caused them to partially open, even though the fluid pressure remained below the remote least principal stress (Fig. 3.3).

The behavior of the injection rate with time was rather complex (Fig. 3.5). Typical for constant pressure injection, the injection rate began high (because there was a large pressure difference between injection pressure and initial pressure) and declined rapidly. The decline in injection rate was reversed as slip began to occur and transmissivity increased around the wellbore. After a period of increase, the injection rate began to fall again as the progressive increase in transmissivity was unable to maintain the flow rate. The spreading of transmissivity enhancement was delayed as the pressure perturbation reached the edges of the central fracture. At around 250 s, the neighboring fractures began to open and slip, increasing their transmissivity and slowing (but not reversing) the decrease in injection rate. Subsequently, injection rate entered a period of gradual decline as the pressure of the entire system slowly approached the injection pressure.

M. W. McClure and R. N. Horne, *Discrete Fracture Network Modeling*   73
of Hydraulic Stimulation, SpringerBriefs in Earth Sciences,
DOI: 10.1007/978-3-319-00383-2_4, © The Author(s) 2013

4.1.2 Effect of Spatial and Temporal Discretization

The effect of spatial and temporal discretization on simulation results is shown in Figs. 3.2, 3.3, 3.4, 3.5, 3.6, 3.7, 3.8. The results were convergent to discretization refinement in both time and space.

Models A1-A4 used constant element size discretizations of increasing refinement. Model A5 used a variable size discretization with significant refinement near the wellbore and fracture intersections. Simulation settings S1-S5 included a variety of values of η_{targ}, the main parameter that controls time step duration.

Figures 3.2 and 3.3 show the final sliding and opening displacements of the simulations with different spatial discretizations (and the COMP2DD result). Other than Model A1, which used the coarsest discretization, results appear quite similar to the most highly refined results. Figure 3.4 shows a calculation of the difference between the various simulations and the COMP2DD result. It shows that the COMP2DD result was virtually identical to the result from Model A4-S3, which used the same discretization. Comparison between A4-S0 and A4-S3 indicates that results with direct BEM multiplication were virtually identical to results with Hmmvp. Figure 3.4 shows that the results were convergent to discretization refinement.

Figures 3.2, 3.3, and 3.4 give some insight into the coarsest discretization that can be reasonably used. Results from A1 were clearly unacceptable, but results from A2 and A5 were reasonably close to the more spatially refined simulations. Comparing between A2 and A5, A5 was modestly more accurate even though it had fewer elements than A2. Evidently, the strategy of refining the discretization around the fracture intersections is better than using a constant element size.

Figures 3.5 and 3.6 show the effect of the spatial discretization on the flow rate history. The A1 simulation was extremely different from the others. The results for A2, A3, and A5 all matched the result from A4 reasonably closely, and A3 was the closest match. A5 was significantly closer to the A4 result than A2 after about 250 s. Overall, every discretization except A1 performed reasonably well in matching the most highly resolved simulation, A4. The A5 simulation was the best compromise between efficiency and accuracy, evidently because it used refinement near the wellbore and fracture intersections.

Figures 3.7 and 3.8 show the effect of time step duration on results. Simulations settings S1-S5 were used with Model A5. The same spatial discretization was used because these simulations were designed to test the effect of temporal discretization. Simulation A5-S1 used an exceptionally small value for η_{targ} and for comparison purposes was assumed to be the most accurate solution. Figure 3.7 shows that A5-S5 was the only result that was extremely different from A5-S1. Simulation A5-S4 showed some difference from the benchmark, and A5-S2 and A5-S3 were very similar to A5-S4. Figure 3.8 shows that the results were convergent to refinement.

The degree of discretization refinement in a simulation must be balanced against efficiency. If the objective of a simulation is to obtain a highly accurate result, a more refined simulation may be required. However, in subsurface modeling, error is usually unavoidable because of poor or incomplete information about the subsurface and simplifying assumptions about the physical processes taking place. Because error due to uncertainty is already high, there little benefit to driving numerical error to nearly zero. Practically, accepting a modest numerical error is justified, especially if a problem is so complex it cannot be solved otherwise. Model A is very simple, and so efficiency was not a major concern, but in larger models, efficiency is critically important. Simulation results with Model A5 were reasonably similar to results from Model A4, yet were roughly 1,000 times more efficient (Figs. 3.4 and 3.6). For this reason, Model A5 was chosen as the best compromise between efficiency and accuracy, and discretization settings similar to A5 were used with Models B and C. For temporal refinement, S3 or S4 were probably the best compromise between efficiency and accuracy (η_{targ} equal to 0.5 and 0.05 MPa, respectively).

4.1.3 Solving Directly for Final Deformations

In simulations A4-S6 and A4-S7, the final deformations of the fractures were calculated in a single step (Sect. 3.2.1). This simulation strategy could be useful if the deformations between the initial and final state were not considered important. In effect, the direct calculation method solves the same contact problem as COMP2DD (Mutlu and Pollard 2008).

Figures 3.10 and 3.11 show that the final displacements from A4-S6 and A4-S7 were virtually identical to the results from COMP2DD (relative error less than 10^{-6}), and the calculations were much more efficient. Comparison between A4-S6 and A4-S7 shows that the direct BEM yielded virtually the same result as Hmmvp, yet the results using Hmmvp were much more efficient.

Simulation A4-S6 (which used Hmmvp) reached virtually the same answer as COMP2DD and was 135 times faster than the Lemke algorithm and 1,550 times faster than the SOCCP algorithm (Fig. 3.9). The time comparisons between COMP2DD and Simulation A4-S6 are not completely equivalent because COMP2DD is a Matlab code and the simulator is written in C++. Typically, Matlab codes have worse performance than C++ code (except for functions that are vectorized in Matlab). Nevertheless, the difference in efficiency cannot be completely explained by the difference between Matlab and C++.

The direct solution method should have excellent scaling with problem size because the computation time is dominated by the matrix multiplications associated with updating stresses, and with Hmmvp, this step scales like n or $n\log(n)$ (Sects. 3.5 and 4.5). Because the model does not ever require assembly of the dense matrix of interaction coefficients (it only requires storage of the highly compressed hierarchical matrix form of the interaction matrices), the RAM requirements are much lower. Therefore, the model could be used to solve extremely large contact problems that would not be feasible for algorithms that require assembly and solution of the full, dense matrix of interaction coefficients.

It is likely that if the model had been more specifically tailored to be used as a direct solution algorithm, the calculation could have been done even more efficiently. The fluid flow equations, while altered to have no effect on the solution, remained in the fluid flow/normal stress system of equations (Sect. 3.2.1), creating unnecessary computational effort. Possibly, the solution could have been made more efficient if full coupling, rather than iterative coupling, had been used.

As discussed in Sect. 2.3.8, it was found that the enforcement of inequality constraints on fracture displacements could lead to convergence failure in the shear stress residual equations for very complex, dense, or poorly discretized fracture systems. Possibly, convergence could be an issue for problems in which a direct solution is attempted because the initial guess may not be close to the final solution. To improve robustness, a sparser iteration matrix could be used (Sect. 2.3.8). A diagonal iteration matrix is the least efficient, but most robust choice. Another alternative would be to apply the stresses gradually over several steps and perform several calculations. An equivalent strategy would be to apply the stresses abruptly, but use time stepping with an artificially large value of the radiation damping coefficient (or equivalently, use a normal radiation damping coefficient but enforce very short time steps), which would force the deformations to occur gradually over several time steps.

4.1.4 Effect of cstress

The effect of the *cstress* option was tested by Simulations A3-(S9-S12). Simulation A3-S9 can be compared to A3-S8, which was identical except that A3-S8 did not use the *cstress* option. The final opening and sliding displacements from these simulations are shown in Figs. 3.11 and 3.12.

Results from A3-S8 and A3-S9 were virtually identical. The value of E_0 in these simulations was very small, and so the closed fractures experienced very little normal deformation and induced little normal stress. As E_0 was increased, from S9 to S12, the normal displacements of the closed fractures became greater, and the induced normal stresses became increasingly significant. As induced normal stresses became greater, the induced opening and sliding of the elements decreased, shown in Figs. 3.11 and 3.12.

These results demonstrate that for small values of E_0 consistent with cracks, the *cstress* option has a limited effect. For larger values of E_0, consistent with faults or fracture zones, the *cstress* option could have a major impact on results. However, these results are probably not realistic (Sect. 2.3.3) because the Shou and Crouch (1995) Displacement Discontinuity Method is probably not appropriate for larger values of E_0. The Shou and Crouch (1995) method is intended to describe the opening of cracks, not the poroelastic swelling of fault zones. Future work is needed to replace the Shou and Crouch (1995) kernel with a boundary element method designed for swelling of porous fault zones.

4.2 Model B

Simulations were performed with Model B to test simulations options on a large, relatively complex fracture network. Model B contained 52,748 elements and 1,080 fractures. Each simulation involved injection at constant rate for 2 h (in simulation time), involved thousands to hundreds of thousands of time steps, and required hours to days of computation time on a single processor.

The fracture network and simulation parameters were not calibrated to match any particular location. Nevertheless, the results could be used to draw some interesting insights about shear stimulation.

4.2.1 General Description of Results

The results from Model B demonstrate pure shear stimulation—transmissivity enhancement was from induced slip on preexisting fractures. Injection pressure was consistently above 50 MPa, the least principal stress (Fig. 3.24), but it was specified as a model parameter that new fractures could not form (Sect. 3.3). As a result, all injected fluid was contained in preexisting fractures, some of which opened as the fluid pressure exceeded their normal stress. The injection pressure was quite variable with time as evolution of transmissivity sometimes caused drops in injection pressure (Fig. 3.24).

All fractures in the model were well oriented to slip in the preexisting stress state, yet not all fractures close to the wellbore slipped (Fig. 3.12). This occurred because of the way that shear stimulation progressed through the network and because of stress shadowing.

The process of Crack-like Shear Stimulation (described in greater detail in Sects. 3.4.2.2 and 4.4.2 of McClure 2012) is very efficient at propagating transmissivity enhancement down a single, linear fracture. An effective shear crack tip develops at the front between where slip has occurred and slip has not yet occurred. The concentration of stress in the neighborhood of the effective crack tip can allow slip to propagate ahead of the front of fluid pressure perturbation. Slip couples to increase in transmissivity, and so the fluid pressure front is able to propagate at a rate related to the stimulated, not the unstimulated, transmissivity. Therefore, once slip begins to occur on a fracture, it tends to relatively rapidly advance down the entire fracture. Figure 4.1 demonstrates the process of Crack-like Shear Stimulation.

Episodic propagation of shear stimulation down faults can be seen in Fig. 3.18, which shows the timing and magnitude of seismic events from Simulation B6.

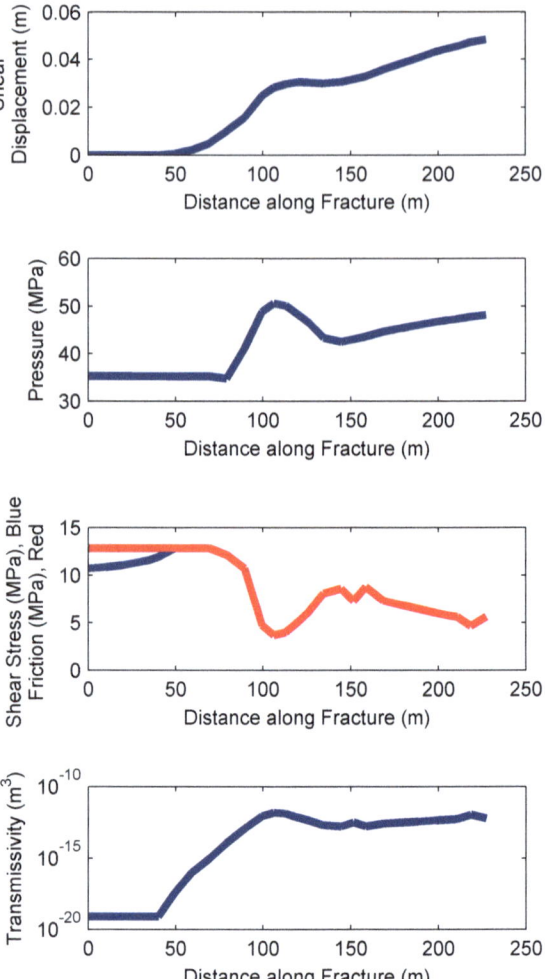

Fig. 4.1 An illustration of the crack-like shear stimulation mechanism. The *x-axis* shows distance along a particular fault. Injection and stimulation are generally propagating *from the right to the left*. The front of fluid pressure perturbation, located around x = 80 m, is lagging behind the front of shear displacement and transmissivity enhancement. Figure from Sect. 3.4.2.2 in McClure (2012)

There were periods of relatively intense seismicity. Each period of seismic intensity corresponded to the progressive advancement of fluid pressure, slip and transmissivity enhancement down a particular fracture. The effects were evident in the plot of injection pressure. For example, there was a period of intense seismicity at around 5,500 s (Fig. 3.18) as slip and transmissivity enhancement advanced down a newly stimulating fracture. At the same time, injection pressure dropped (Fig. 3.25). There was a similar, smaller drop in injection pressure that corresponded with a period of intense seismicity around 5,000 s (Figs. 3.18 and 3.25).

In contrast to propagation of slip down a fracture, the initiation of slip on a fracture is (typically) not aided by induced stresses. Most commonly, initiation of slip requires fluid pressure to diffuse into the (unstimulated) fracture, which is a process that is rate-limited by the initial transmissivity of the fracture (see the discussion in Sect. 3.4.2.2 of McClure 2012).

When fractures slip, they relieve shear stress along their sides, inhibiting neighboring, parallel fractures from also slipping, a process referred to as stress shadowing. In the simulations, the spatial range of the perturbation was limited by the height of the formation, 100 m, because the Olson (2004) correction was used (Sect. 2.3.2). Stress interaction was manifested in other ways. Fracture opening occurred in the extensional quadrants of slipping fractures (Fig. 3.12) due to induced reductions in normal stress.

These effects explain why many apparently well oriented fractures did not slip. The most optimally oriented fractures intersecting the wellbore slipped earliest. Once they began to slip, the shear stimulation process allowed stimulation to propagate along them quite rapidly. The fractures that slipped first caused stress shadows that prevented or delayed the neighboring fractures from slipping.

The final distribution of transmissivity was relatively extensive spatially, spreading 200 m from the wellbore. Yet most connections to the wellbore were relatively isolated from each other. There were not necessarily direct flow pathways between neighboring stimulated fractures. This is a realistic result for flow in fracture networks in low permeability matrix, and it occurred in the model as a direct consequence of the use of discrete fracture network modeling and the inclusion of stresses induced by fracture deformation.

4.2.2 Effect of Temporal Discretization Refinement

Simulations B1-B4 tested the performance and accuracy of the model at various values of η_{targ}, the parameter that determines time step duration (Sect. 2.3.9). Figures 3.12 and 3.13 show the final sliding and opening displacements for Simulations B1 and B4, the most and least temporally resolved simulations. With visual inspection, small differences can be observed (most notably, there is an entire fracture that slipped in Simulation B4 that did not slip in B1), but the results were generally rather similar. Plots of the final displacements for B2 and B3 are not shown because they are visually indistinguishable from the results from B1 in Fig. 3.12.

From these results, it can be concluded that the value of η_{targ} in Simulation B4 was too large, but the value for η_{targ} in B2 and B3 was sufficient for reasonably accurate results. Figure 3.14 demonstrates that results the results were convergent to temporal refinement. The behavior of injection pressure with time was visually identical for Simulations B1-B3. Simulation B4 was similar, but a major difference was that there was a drop in injection pressure near the end of the simulation that did not occur in Simulations B1-B3 (Fig. 3.24). The drop in injection pressure occurred when shear stimulation rapidly propagated a particular large fracture. This fracture was stimulated in Simulation B4 but not in Simulations B1, B2, and B3 (Figs. 3.12 and 3.13), and so the drop in injection pressure did not occur in those simulations.

Run-time and number of iterations increased significantly as η_{targ} decreased (Fig. 3.21). With larger η_{targ}, average time step duration increased, and so the iterative coupling scheme required more iterations between the shear stress residual equations and the flow/normal stress residual simulations (see Sect. 2.3.1). As a result, the simulations with larger η_{targ} took fewer time steps but had greater run-time per time step (Fig. 3.22).

4.2.3 Effect of cstress

Simulations B8 and B9 tested the *cstress* option with an E_0 equal to 0.8 mm. Testing of *cstress* with Model A3-(S9-S12) suggested that with E_0 equal to 0.8 mm, the *cstress* option would probably have a moderate effect on the results. For example, Simulation A3-S11, which used E_0 equal to 1 mm, was significantly affected by the *cstress* option (Figs. 3.10 and 3.11).

From visual inspection, the plot of final slip and opening for Simulation B8 (Fig. 3.17) appears reasonably similar to the result from Simulation B1 (Fig. 3.12). The *cstress* option may have had less effect in Model B because the fractures were significantly larger than in Model A. Because fracture stiffness decreases with length, less stress is induced on a fracture of greater length for the same normal displacement. Figure 3.14 shows that within the stimulated fractures, the Euclidean norm (scaled for problem size) of the difference in sliding displacement between Simulation B8 and B1 was roughly 4 mm. This difference is fairly significant because the average sliding displacement in the stimulated region was about 2.3 cm.

The number of time steps performed during B8 simulations was almost identical to the number performed in B3 (which used the same η_{targ}), but the simulation run-time was 5.2 times greater. Computation time per time step was roughly 5.9 s for B8 and 0.8 s for B3. The additional simulation time was needed for two reasons. With the *cstress* option, updating the stresses caused by deformation of the closed elements requires multiplication by the matrix of interaction coefficients (using Hmmvp). Second, the iteration matrix in the flow/normal stress system of equations is significantly larger and requires more time to solve (Eq. 2.30).

4.2.4 Effect of Adaptive Domain Adjustment

Simulations B5 and B9 used adaptive domain adjustment. These simulations can be compared to B3 and B8, which were otherwise identical to B5 and B9 but did not use adaptive domain adjustment. The reduction in computation time achieved by the adaptive domain adjustment was around 15 % for the simulations not using *cstress* and about 30 % for the simulations using *cstress* (Figs. 3.21 and 3.23). The simulation results using adaptive domain adjustment were virtually identical to the simulations not using adaptive domain adjustment.

From Fig. 3.23, it can be seen that at early time the simulations using adaptive domain adjustment were significantly more efficient, but after the early stages of the simulation, the advantage disappeared. It was expected that the efficiency gain would only occur early in the simulation because as pressure perturbation and simulation spread from the injector well, the number of elements included in the *checklist* increased until it included all elements.

4.2.5 Dynamic Friction Weakening

Simulation B6 was performed to test the effect of dynamic friction weakening (Sect. 2.5.3). Seismic event magnitudes spanned a range between -2.5 and 2.0 (Fig. 3.18). The minimum magnitude was determined by the minimum element size in the model. Because slip was typically confined to a single fracture during a seismic event, the maximum magnitude was limited by the largest fracture in the model.

Due to the Crack-like Shear Stimulation mechanism (Sect. 4.2.1), there were periods of relatively intense seismicity separated by periods of relatively mild seismicity (Fig. 3.18). The periods of greater activity occurred as slip and transmissivity enhancement propagated down a fracture for the first time. Once a fracture began to slip, transmissivity enhancement and fluid pressure were able to propagate down the fracture relatively quickly because slips induced shear stress along the fracture, which encouraged slip and transmissivity enhancement.

Figure 3.20 shows the locations of the hypocenters with an adjustment to simulate the effect of relocation error. The relocation causes the hypocenters to form a volumetric cloud. However, Fig. 3.15 shows the final slip distribution (with transmissivity enhancement closely correlated to slip), and it demonstrates that the actual fracture network is relatively sparse, widely spaced, and poorly connected. This result is consistent with results from EGS, for example, where production logs sometimes demonstrate a wide spacing of flowing fractures yet microseismic relocations appear volumetric (Michelet and Toksöz 2007).

The results indicate the peril of attempting to infer fracture geometry from the shape of a microseismic cloud. Without knowing the actual locations of stimulated fractures (Fig. 3.15), an observer might infer from Fig. 3.20 that there is a long fracture oriented in the direction of the y-axis at the location $x = -150$ m. Figure 3.15 shows that no such fracture exists.

Simulation B6 was slightly more efficient per time step than Simulation B3, which used the same η_{targ}. Simulation B6 used significantly more time steps than Simulation B3 because a large number of short time steps were needed to simulate the rapid slip during seismic events.

4.2.6 Neglecting Stress Interaction

In Simulation B7, neglecting stress interaction between fractures caused the stimulated fracture geometry to be completely different (Fig. 3.16). Stress interaction is rarely included in models of large, complex fracture networks (Sect. 1.2), but this example demonstrates how profoundly stress interaction affects simulation results. There are many effects from stress interaction, and one of the most important is the Crack-like Shear Stimulation mechanism (Sect. 4.2.1). Induced shear stresses help stimulation propagate along fractures. Without that effect, the stimulated region (in shear stimulation) can only grow as rapidly as flow can occur into elements at initial transmissivity, a process that can be very slow if initial transmissivity is low. If the region of transmissivity enhancement grows too slowly to accommodate the injected fluid, fluid pressure is forced to rise, causing opening.

In Simulation B7, new fractures were not permitted to form (Sect. 3.3), but preexisting fractures were permitted to open. The thickness of the stimulated fractures in Fig. 3.16 indicates that these fractures have opened significantly. The crack tip region adjustment (Sect. 4.2) was not deactivated in Simulation B7, and this process turned out to be the primary way that the region of stimulation was able to grow. Figure 3.25 shows that the injection pressure during Simulation B7 was almost constant at around 55 MPa. This was the injection pressure at which the natural fractures were able to open, enabling the crack tip adjustment to be activated.

4.3 Model C

Model C is an example of mixed-mechanism propagation, where stimulation progresses through both propagation of new fractures and opening and sliding of preexisting fractures (Sect. 3.1.1 in McClure 2012). Mixed-mechanism propagation is most commonly proposed as an explanation for gas shale stimulations (Gale et al. 2007; Weng et al. 2011). The natural fracture network was not percolating—there were no continuous pathways through the reservoir contained only in the natural fractures However, continuous, long distance pathways for flow were able to develop because of the formation of new fractures (Fig. 3.26). This is an important point of distinction between pure shear stimulation and mixed-mechanism propagation. Shear stimulation requires percolation of the natural fracture network while mixed-mechanism propagation does not (Sect. 3.4.8 in McClure 2012).

Model C demonstrated that the model is capable of performing efficiently for mixed-mechanism stimulations. Model C ran significantly faster and used fewer time steps than Model B3, which used the same value for η_{targ}. However, Model C had fewer than half as many elements as Model B3 (Sect. 3.1).

4.4 Model D

The Model D simulations were performed to test the efficacy of the strain penalty factor and to investigate model behavior at low-angle fracture intersections. In the two simulations without the strain penalty factor, there were unrealistic numerical artifacts, with rapid oscillation between large and small opening between elements (Figs. 3.27 and 3.29).

Comparison between D1-DS1 and D2-DS1 shows that discretization refinement reduces numerical artifacts at low-angle intersections, but it cannot eliminate then completely. In D2-DS1, the region of numerical artifacts is smaller, localized in the immediate vicinity of the intersection, while in D1-DS1, numerical artifacts are present a significant distance from the intersection.

Numerical artifacts such as oscillations in fracture opening are not present in the simulations that used the strain penalty method, D1-DS2 and D2-DS2 (Figs. 3.28 and 3.30). In D2-DS2, modest penalty stresses applied very near the intersection prevented numerical oscillations without having a major effect on the overall results. In D1-DS2, penalty stresses prevented numerical oscillations, but comparison to D2-DS2 shows that the results were affected significantly. These results show that while the penalty stress method prevents numerical artifacts, it also causes inaccuracy. D2-DS2 shows that the inaccuracy caused by the penalty method can be limited to a very small region if the discretization is reasonably refined around the intersection.

4.5 Hierarchical Matrix Decomposition

Hmmvp demonstrated exceptional performance and scaling for matrix multiplication. For progressive refinement of a fracture network, Hmmvp had linear scaling with problem size (Fig. 3.32). For constant level of refinement and increasing problem size, Hmmvp had $n\log(n)$ scaling with problem size (Fig. 3.34). For creation of the matrix approximation, Hmmvp had linear scaling (Fig. 3.35).

Tests of Hmmvp with the simulator demonstrated its accuracy and efficiency. The settings for Simulations A4-S0 and A4-S3 were identical except that A4-S0 used direct matrix multiplication instead of Hmmvp. The results were virtually identical, yet A4-S0 required roughly ten times more computation time (Fig. 3.4). A similar comparison was made between Simulations A4-S6 and A4-S7, and in this case, the simulation using Hmmvp was 23 times more efficient. The A4-S6 and A4-S7 simulation results were virtually identical (Fig. 3.9). Because FLOPs per multiplication achieved by Hmmvp grows much more slowly with size than direct matrix multiplication, the efficiency gain grows as the problem size increases.

4.6 Extension of the Model to Three Dimensions

The model described in this book could be extended to three-dimensional modeling. Stress calculations would be identical except that a different boundary element method would be used to calculate interaction coefficients. In the fluid flow equations, three-dimensionality would only change the calculation of the geometric transmissibility term. A few specific details would be changed, such as changing the way that the stress intensity factors are calculated (Sect. 2.5.2). Changes would also be required for discretization and visualization of results.

There are several physical issues specific to three-dimensional models that could be challenging to handle due to their inherent complexity. For example, a propagating mode I fracture subjected to mode III loading may form an en echelon array of smaller fractures (Pollard et al. 1982).

The main challenge of three-dimensional simulations is that they would require a greater number of elements. Because our model only requires discretization of the fractures, the number of elements would be much smaller than if volumetric discretization was required (such as with finite element). Because the Model B simulations ran in hours or days on a single computer, parallelization would be necessary for simulating substantially larger problems. Fortunately, there is no major theoretical obstacle to parallelization. The matrix multiplication aspect of the problem would be trivial to parallelize and has excellent scaling with problem size. Solving the iteration matrix system in the flow/normal stress subloop could be done with standard parallel solvers for sparse systems.

References

Gale, J.F.W., Reed, R.M., Holder, J.: Natural fractures in the Barnett Shale and their importance for hydraulic fracture treatments. AAPG Bull. **91**(4), 603–622 (2007). doi:10.1306/11010606061

McClure, M.W.: Modeling and characterization of hydraulic stimulation and induced seismicity in geothermal and shale gas reservoirs. Stanford University, Stanford (2012)

Michelet, S., Toksöz, M.N.: Fracture mapping in the Soultz-sous-Forêts geothermal field using microearthquake locations. J. Geophys. Res. **112**(B7), (2007). doi:10.1029/2006JB004442

Mutlu, O., Pollard, D.D.: On the patterns of wing cracks along an outcrop scale flaw: a numerical modeling approach using complementarity. J. Geophys. Res. **113**(B6), (2008). doi:10.1029/2007JB005284

Olson, J.E.: Predicting fracture swarms—the influence of subcritical crack growth and the crack-tip process zone on joint spacing in rock. Geol. Soc., Lond. Special Publ. **231**(1), 73–88 (2004). doi:10.1144/GSL.SP.2004.231.01.05

Pollard, D.D., Segall, P., Delaney, P.T.: Formation and interpretation of dilatant echelon cracks. GSA Bull. **93**(12), 1291–1303 (1982). doi:10.1130/0016-7606(1982)93<1291:FAIODE>2.0.CO;2

Shou, K.J., Crouch, S.L.: A higher order Displacement Discontinuity Method for analysis of crack problems. Int. J. Rock Mech. Min. Sci. Geomech. Abstr. **32**(1), 49–55 (1995). doi:10.1016/0148-9062(94)00016-V

Weng, X., Kresse, O., Cohen, C.-E., Wu, R., Gu, H.: Modeling of hydraulic-fracture-network propagation in a naturally fractured formation. SPE Prod. Oper. **26**(4), (2011). doi:10.2118/140253-PA

Chapter 5
Conclusions

The modeling methodology described and demonstrated in this book is capable of efficient and accurate simulation of fluid flow, deformation, seismicity, and transmissivity evolution in large two-dimensional discrete fracture networks. Appropriate stress conditions and constraints on displacements are applied on elements depending on whether they are open, sliding, or stationary. Results are convergent to grid refinement, and discretization settings required for acceptable accuracy were identified. A variety of techniques that enable efficiency and realistic model behavior—such as adaptive domain adjustment, crack trip region adjustment, and the strain penalty method—have been developed and tested. The model can be used for direct solution of fracture contact problems in a way that has minimal memory requirement, excellent efficiency, and desirable scaling with problem size.

Test results demonstrated the critical importance of including stresses induced by deformation in modeling of stimulation. These stresses directly impact the mechanism of stimulation propagation and the properties of the resulting fracture network.

The model can be used to explore the behavior of hydraulic stimulation in settings where preexisting fractures play an important role. The model can be used to describe propagation of new opening mode fractures, induced slip on preexisting fractures, or any combination of those processes.

M. W. McClure and R. N. Horne, *Discrete Fracture Network Modeling of Hydraulic Stimulation*, SpringerBriefs in Earth Sciences, DOI: 10.1007/978-3-319-00383-2_5, © The Author(s) 2013

List of Variables

A	Fracture surface area, m^2
activelist	List of elements maintained for use in adaptive domain adjustment
a	Element half-length, m
a_{const}	Element half-length for initial discretization, m
a_{frac}	Fracture half-length, m
a_{min}	Minimum element half-length, m
a_{rs}	Rate-state friction, coefficient for velocity term, unitless
$B_{E,\sigma}$, $B_{D,\sigma}$, $B_{E,\tau}$, $B_{D,\tau}$	Matrices of interaction coefficients, MPa/mm
b	Rate-state friction, coefficient for state term, unitless
checklist	List maintained of elements that are fully included in the problem domain with adaptive domain adjustment
cstress	An simulation option to include the stress caused by normal displacement of a closed element
D	Cumulative shear displacement discontinuity, mm
D_{cum}	Cumulative rapid slip during a seismic event, mm
$D_{E,eff}$, $D_{e,eff}$	Effective cumulative displacement discontinuity, mm
$D_{e,eff,max}$, $D_{E,eff,max}$	Maximum effective cumulative sliding displacement used for calculating aperture, mm
D_s, D_n	Variables used in the description of the stress penalty method, referring to shear and normal displacement discontinuity (equivalent to D and E), mm
d	Distance between the centers of two elements, m
d_c	Rate-state friction, characteristic weakening distance, m
dt	Duration of a time step, s
E	Void aperture, mm
E_0	Reference void aperture, mm
E_{hfres}	Residual closed aperture, mm
E_{open}	Open aperture, physical separation between walls, mm
e	Hydraulic aperture, mm
e_0	Reference hydraulic aperture, mm
e_{hmat}	Relative error for h-matrix approximation, unitless

M. W. McClure and R. N. Horne, *Discrete Fracture Network Modeling of Hydraulic Stimulation*, SpringerBriefs in Earth Sciences, DOI: 10.1007/978-3-319-00383-2, © The Author(s) 2013

e_j	Relative difference between simulations, defined in a variety of ways
e_{proc}	Process zone hydraulic aperture, mm
f_0	Rate and state friction term, unitless
G	Shear modulus, GPa
G_{adj}	Olson (2004) adjustment factor to interaction coefficients, unitless
h	Out of plane fracture width, or height, m
I	Unit matrix, unitless
$itertol$	Convergence tolerance for iterative coupling, MPa
J	Iteration matrix, various forms
$J_{mech,thresh}$	Threshold parameter for including mechanical interaction terms in the iteration matrix, unitless
K_I	Stress intensity factor, MPa-m$^{1/2}$
$K_{I,crithf}$	Critical stress intensity factor for propagation of a new fracture, MPa-m$^{1/2}$
$K_{I,crit}$	Critical stress intensity factor for propagation of opening on a preexisting fracture, MPa-m$^{1/2}$
K_{frac}	Fracture stiffness, MPa^{-1}
K_{hf}	Stiffness of closed, newly formed fractures, MPa^{-1}
k	Permeability, m^2
l_c, l_o, l_s, l_f	Parameters used for discretization refinement, m, unitless, unitless, and unitless
M_0	Seismic moment, N-m
M_w	Moment magnitude, unitless
$mechtol$	Convergence tolerance for the shear stress residual equations, MPa
$nochecklist$	List maintained of elements that are not fully included in the problem domain with adaptive domain adjustment
$nocstress$	An simulation option not to include the stress caused by normal displacement of a closed element
$opentol$	Tolerance for including elements with adaptive domain adjustment, MPa
P	Pressure, MPa
P_{init}	Initial formation fluid pressure MPa
P_{inj}	Injection pressure, MPa
$P_{prodmin}$	Minimum production pressure, MPa
P_{injmax}	Maximum injection pressure, MPa
Q	Calculated injection rate, kg/s
q	Mass flow rate, kg/s
q_{flux}	Mass flux, kg/(s-m^2)
q_{injmax}	Maximum injection rate, kg/s
$q_{prodmax}$	Maximum production rate, kg/s

R	Residual equation, various units
S	Specified total injection rate, kg/s
S_0	Cohesion, MPa
$S_{0,open}$	Cohesion term for open elements, MPa
s_a	Mass source term per area, kg/(s-m^2)
s	Mass source term, kg/s
$slidetol$	Tolerance for including elements with adaptive domain adjustment, MPa
T	Transmissivity, m^3
$T_{hf,fac}$	Factor for calculating residual transmissivity of newly formed fractures, m^2
T_g	Geometric transmissibility between elements, m^3
T_s	Stress tensor, MPa
t	Time, s
v	Sliding velocity, m/s
v_0	Rate-state friction, reference velocity, m/s
v_s	Shear wave velocity, m/s
X	Vector of unknowns, various forms
x	Dummy variable to specify an arbitrary direction, m
$\Delta D, \Delta E$	Change in displacement discontinuity during a time step, mm
$\Delta\sigma_{n,strainadj}, \Delta\sigma_{s,strainadj}$	Stress (normal or shear) applied in the strain penalty method, MPa
δ	Parameter used in adaptive time stepping, sum of absolute value of change in shear stress and effective normal stress, MPa
$\delta_{strainadj}$	The largest value of $\Delta\sigma_{k,strainadj}$ during a time step, MPa
ε	Strain tensor, unitless
$\varepsilon_n, \varepsilon_s$	Displacement discontinuity normal or shear strain, used for high strain penalty method, unitless
$\varepsilon_{n,lim}, \varepsilon_{s,lim}$	Limit to displacement discontinuity strain (normal or shear), unitless
ε_{tol}	User specified relative tolerance for h-matrix assembly, unitless
η	Radiation damping coefficient, MPa/(m/s)
η_{targ}	Target change in stress change parameter for adaptive time stepping, MPa
$\eta_{targ,strainadj}$	Target change in ε_k for time stepping, unitless
θ	Rate-state friction, state variable, s
μ_d	Dynamic coefficient of friction, unitless
μ_f	Coefficient of friction, unitless
μ_l	Fluid viscosity, Pa-s
ρ	Density, kg/m^3
π	Mathematical constant Pi, unitless

σ_n'	Effective normal stress, MPa
σ_n	Normal stress, MPa
$\sigma_{n,Eref}$, $\sigma_{n,eref}$	Reference fracture stiffness, mm
σ_{yy}	Remote compressive stress in the y direction, MPa
σ_{xy}	Remote shear stress, MPa
σ_{xx}	Remote compressive stress in the x direction, MPa
τ	Shear stress, MPa
υ_p	Poisson's ratio, unitless
$\varphi_{E,dil}$, $\varphi_{e,dil}$	Aperture dilation angle, °